MORE FOOD FROM SOIL SCIENCE

By V. A. TIEDJENS

Encyclopedia of Vegetable Gardening
More Food From Soil Science

COAUTHOR OF

Chemical Gardening for the Amateur
Handbook of Gardening
A Practical Guide to Successful Farming

MORE FOOD FROM SOIL SCIENCE

The Natural Chemistry of Lime in Agriculture

V.A. TIEDJENS

First Printing, MCMLXV
Second Printing, MCMLXXVII
Third Printing, MCMLXXXIX
Fourth Printing, MCMLXXXXI
Fifth Printing, MCMLXXXXIV

Growers Chemical Corp.
Milan, Ohio 44846

Library of Congress Catalog Card Number: 65-28079

ISBN 0-682-43057-9

Printed in the United States of America

MORE FOOD FROM SOIL SCIENCE

Preface

JESUS' FEEDING the Philistines with five loaves of bread and two fishes was considered a miracle. If the Bible is true to facts, this gives us something to think about. However, to grow 200 bushels of corn where only 65 bushels grew before is no less a miracle, because it means that we can feed three times as many people. This yield can mean life or death for millions of earth's inhabitants, and can postpone the day of reckoning for several generations.

There has been a tremendous change in life on the farm since the early 1900's. The transition from the backwoods manner of 1900 to today's swanky farm homes, with conveniences comparable to the best our city cousins enjoy, has been correlated closely with the managerial ability of owners: it has depended on how much profit is made from the soil.

Generally speaking, in every area, regardless of existing soil conditions, there are examples of both good and poor management—whether the soil is fertile or submarginal (submarginal, as some of our experts are ever ready to classify it). This is encouraging, because it demonstrates that there are ways and means of growing good crops on even our poorest soils—that it is within the power of man to treat even the poorest soils in such a way as to make farming them a profitable venture. And it is possible to select a good location—where markets, schools and transportational facilities are established—by buying a "worn-out" farm. Then all a person needs is the know-how to build those worn-out acres into highly productive land.

There are many things that a farmer can do to grow more bushels on an acre of ground. A number of problems confront a farmer. Some are under his control; over others he has no control. He probably has little control over the weather and the elements —such things as early autumn and late spring frosts can be guarded against, but only at an expense that precludes profit. Average summer temperatures cannot be changed, except in localized areas at considerable expense. There is no control over high winds, hailstorms, excessive heat, cloudy weather, and excessive rainfall. Irrigation, where water is available, can offset the effect of a severe drought; but here again, cost comes into the picture. Cultural practices are at the discretion of the farmer. Cultivating the soil, including plowing and fitting, are again within the scope of man's control and judgment. This has a bearing on those factors over which he has very little control.

Good well water is very important to a location. Markets are extremely important and must be considered when deciding on a location. Proximity to schools, churches and stores is a factor over which man has some control. There are a large number of factors over which a farmer has good control if he understands how his soils and crops function. These he must be held responsible for. They are the factors that will probably influence his yields the most, and the ones that will be discussed here at greatest length.

Cultivating the soil is one of many vocations open to our young people. Living on a farm involves our health, our pleasures, our finances. For persons who dislike farming, it can be drudgery; for others, it can be a wonderful life. But farming is a business. It behooves us to choose a profession we enjoy.

"Calcium to riches" may seem farfetched when applied to the agricultural industry, but as a result of my experience I have formulated a hypothesis which in fact makes calcium the keystone to success in the art of growing food crops.

There are many books and magazines which carry discussions of crop surpluses in some areas and crop shortages in others where agricultural frontiers are still in existence. Some project ideas that can postpone fulfillment of the Malthusian doctrine for years to come, if not forever. Many of these proposals are not practical, even though we have many agricultural readers who can talk glibly about them.

One of the farm surpluses that has given us the most trouble has been that of farm leaders. Conversely, the one big deficiency that still confronts us is a lack of fundamental information on how to maintain high yield potentials in farm soils.—LEGGE

We have the soil, we have the implements, we have suitable weather and the means of supplementing it to grow crops; but with all our scientific knowledge we lack the practical know-how that will unlock the billions of tons of plant food materials and make them available for growing food and fiber plants.

References used freely in formulating a program for high yields are listed below for the convenience of students who, as a result of reading my discussion, may be encouraged to become familiar with the scientific background. I am including them informally because of the nature of this book. This list includes only a few of the many references which should be studied to gain a broader knowledge of the basis of soil fertility. In recent years many papers have been published that form a solid foundation for the program of crop production as presented in this layman's discussion.

GANS, R. Zeolithe und ähnliche Verbindungen ihre Konstitution und Bedeutung für Technik und Landwirtschaft, *Jahr. Preuss. Landesanst. Bergakad.*, 26:179–211 (1905).

________. Konstitution der Zeolithe ihre Herstellung und technische Verwendung, *Jahr. Preuss. Landesanst. Bergakad.*, 27:63–94 (1906).

GEDROIZ, K. K. Soils unsaturated with bases. Method for determining in soils the hydrogen present in an absorbed condition. Soil requirements of lime as a neutralizing agent, *Zür. Opit. Agron.*, 22:3–37 (1924).

________. Ultramechanical composition of soils and its dependence on the nature of cations present in the soil in an absorbed condition. Liming as a means of improving the ultramechanical composition of the soil, *Zür. Opit. Agron.*, 22:29–50 (1924).

________. Exchangeable cations of the soil and the plant. (1) Relation of plant to certain cations fully saturating the soil exchange capacity, *Soil Sci.*, 32:51–63 (1931).

HISSINK, D. J. Beitrag zur Kenntnis der adsorption vorgänge im Boden. Methode zur bestimming der anstauschfähigen oder adsorbtiv gebundenen Basen im Boden und die Bedeutung diesen Basen für die Prozesse die sich im Boden abspielen, *Internat. Mitt. Bodenk.*, 12:81–172, 1922. Base exchange in soils (translation), *Faraday Soc.*, 20:551–66 (1925).

KELLEY, W. P. A general discussion of base exchange in soils, *Jour. Amer. Soc. Agron.*, 18:450–58 (1926).

MATTSON, SANTE. The relation between the electrokinetic behavior and the base exchange capacity of soil colloids, *Jour. Amer. Soc. Agron.*, 18:450–58 (1926).

MACINTIRE, W. H. Reciprocal repression by calcic and magnesic additions in surface soil, *Jour. Amer. Soc. Agron.*, 18:482–97 (1926).

MARSHALL, C. EDMUND. The Colloid Chemistry of the Silicate Minerals, Vol. I. New York: Amer. Soc. Agron. Acad. Press, 1949.

PARKER, FRANK W. Base exchange in soil colloids and the availability of exchangeable calcium in different soils, *Jour. Amer. Soc. Agron.*, 18:470–82 (1926).

TRUOG, E. The cause and nature of soil acidity with special regard to colloids and absorption, *Jour. Phys. Chem.*, 20:457–84 (1916).

WAY, J. THOMAS. On the power of soils to absorb manure, *Jour. Royal Agr. Soc. England*, 11:313–79 (1850).

U.S.D.A. Yearbook, *Soils and Men*. 1938.

U.S.D.A. Yearbook, *Soil*, 1957.

I appreciate having had the privilege of reading these publications, which cover scientific findings up to the first of January, 1963.

Many other papers have been published, but because of the nature of the present book, it seemed unwise to list them all.

V. A. T.

Contents

Introduction

THIS BOOK is written for general information, particularly for the person who is interested in growing crops more profitably on already cultivated soils. It has wide application and can be helpful in most areas where food crops can be grown. The ideas expressed in these chapters are based on sound fundamental information. They can be demonstrated under field conditions. They are controversial when compared with much of the knowledge depended on today to grow our world food supply; but they are not controversial to the research man who knows all the literature pertaining to the subjects discussed.

Growing more food per acre at a lower unit cost, especially in the face of ever-increasing labor costs, should be the goal of every person concerned with the welfare of the farmer. Accumulating food surpluses on the American continent have lulled us into a complacency which has dulled our thinking and may return the agricultural industry to the days of Plato. "If we have too much, why worry about the future?"

"The illusion that times that were are better than times that are has probably pervaded all the ages" (Horace Greeley). We can go back six thousand years to the Prisse Papyrus and read, "Alas, times are not what they used to be." But there are many urgent things to do. We seem to think in circles. Unknowingly we are standing still.

We have been spending our time thinking about whose back we should pat at the next convention, rather than scrutinizing the context of the deeds for which we want to pat someone on the back. Our research thinking has been dull and listless. Our research on food production has been too spasmodic, probably because our research contributions have had to be gleaned with a fine-tooth comb. During the past seventy-five years of soils and crops research we have gradually gleaned some information that has helped to improve our understanding of the physiology of crop production and the chemical reactions taking place in cultivated soil; but this has not resulted in worthwhile yield increases. During that same period we have formulated many hypotheses and theories and have speculated on how we will feed the multitudes in the future. For some reason we have barely dented the surface. During my lifetime we gradually increased our average yield of corn 20 bushels per acre. So we throw out our chests, pat somebody on the back, and congratulate him for having contributed to that increase; but when we look further we find that this accomplishment was brought about only because Providence smiled and provided the growing conditions that made it possible.

When we look for the reasons for the increase of 20 bushels, we find various groups with different interests taking the credit. The group with the biggest political lobby, the commercial fertilizer industry, likes to take most of the credit, when actually they have contributed the least and have probably done the best job of confusing our thinking. Because of the influence of the lobby, many of our young scientists have been dazzled into thinking that propaganda is truth, to the extent that they have allowed this propaganda to guide their thinking. When a person thinks one way for long enough, he begins to adapt his thinking into a lifetime philosophy which he is more and more reluctant to change, even if it is faulty.

The fertilizer industry has been guided by agronomists who got their knowledge from the fertilizer industry. This is because our experiment stations have not had sufficient funds to carry out their programs. When the industry forced its demands for information, the industry was urged to help support the work. It, in good faith, made grants for research fellowships which have helped many college graduates do enough work to earn higher degrees. But when you view this practice in terms of its results, you can't help but think of one of our World War songs, "Don't Bite the Hand That Is Feeding You."

I have had considerable experience with such grants. I received a grant from a large chemical company to study the utilization of nitrogen in plants. I became convinced that much of our so-called factual knowledge gained from research stems from a faulty premise. Since my results did not sell more fertilizer, the fellowship was quickly discontinued.

In later years I accepted a fellowship from a company selling potash. I picked a student well trained in chemistry to work under the conditions of the fellowship. He was very conscientious and did an excellent job on the role of the potassium ion in the growth of plants. He published four worthwhile papers on the subject. But, again, because his work did not help to sell more potash, my fellowship was transferred to another college. My graduate student had to finish his work with no financial assistance. He was the best-informed student I have ever known. He would have made many contributions to our knowledge of the use of plant food materials; but his work was not what the sales manager wanted from the research. The student was not popular with industrial people because he was too sincere and said what he thought. Most people who refused to hire him gave the excuse that he was a Jew. I finally helped place him in the United States Department of Agriculture, where he is doing a fine research job.

If the fertilizer industry has no right to claim credit for the increased yield in corn, we must look further. What I say is based only on my experience and my reading of published reports. Hybrid varieties probably contributed 10 bushels, more or less, to corn yields, partly because they reduced disease and produced better stands with more uniform ear and stalk growth. Planting more seed to the acre added several bushels. Weed control with weed killers added 3 to 5 bushels. Agricultural practices could also have added a few—so we can account for the 20 bushels without giving commercial fertilizer any credit. As a matter of fact, if the real reason were known, we would probably find that we are able to grow an average yield of around 68 bushels of corn without any fertilizer, because records of field plots show that the use of limestone probably deserves more credit than fertilizer. That fertilizer does not deserve the credit is no criticism of commercial fertilizer; it is a criticism of the men who recommend its use. Fundamental information has been lacking to evaluate the need of adding commercial fertilizer.

Fertilizer, from the inception of its use for crop production, was recommended with no reference to the lime condition of the soil. As a result, many tons of fertilizer have been wasted. I want to emphasize this statement, because there is a tendency for writers of future food predictions to take refuge in the idea that we have unlimited fertilizer resources, which will provide for thousands of years hence. Actually, from my own experience, I would feel very much concerned if I thought that our future food supply depended only on our unlimited supplies of fertilizer.

If the ideas we have had for increasing yields in the past seventy-five years had been valid, we would be growing 300 bushels per acre of corn today; but there is no merit to those ideas. I can remember when the soil acidity test was proposed at one of our world soil conferences as the "answer to a maiden's prayer." It was the crystallization into one simple test of many previous ideas, but it dealt with acids and alkalies rather than with limestone and fertilizer residues. When it was used on soil which had not been contaminated by additional chemicals it gave us a valuable research tool. But the use of chemicals for crop production introduced a factor which many failed to take into consideration. We were testing more than the calcium ion, and we began to fall short in our limestone applications. We must keep in mind that even though natural phenomena may have a simple explanation, it may take the combination of many brains to deduce a workable hypothesis. The pH test was reliable but our interpretation led us astray. A soil acidity test gave us a balance sheet on all plus and minus charges in the soil, but since the plus charges did not coincide with the calcium ions (and it was the number of calcium ions we were interested in) the results of the test did not always result in better crop yields. The pH was not as useful as everyone expected. Unfortunately, most of our limestone needs are still being measured with this acidity test; and the addition of anhydrous ammonia to the soil has overshadowed the effect of limestone and has resulted in many alkaline readings which were not due to limestone.

"Oxidation-reduction potential," a rather euphonious phrase, was all the rage at another soil congress. Actually this idea had more potential value for helping raise crop yields, if properly interpreted. Too few realized it merely meant good or poor drainage. Good oxidation meant better root growth and, therefore, better yields. But it, too, was a disappointment, because it was not considered in relation to the other factors affecting plant growth.

Then along came profiles—soil profiles—which were another tool to help increase yields. Much fundamental information was needed to interpret what we saw. This necessitated a thorough knowledge of soil chemistry which few agronomists had, so the presence of a bad plow soil or the physical condition of the soil in the different horizons of the profile could not be translated into crop yields. And yet, in the hands of a soil chemist, a study of the profile made it possible to predict future yields. Dr. Jacob Joffee, formerly with the New Jersey Experiment Station, became very proficient at estimating future yields through the study of soil profiles. I had the pleasure of working with him for many years. In one project we studied soil profiles on more than one hundred farms for three years and estimated possible yields of tomatoes on the appearance, odor and compaction of the soil. I was amazed to find that he had estimated the yield correctly 84 per cent of the time.

I doubt whether we can say that these big yields we hear about are the result of planned treatment. It is true that many of them have received heavy amounts of plant food; but who can say that we might not have had larger yields if considerably less plant food and more limestone had been applied? For advertising fodder, fertilizer companies have used high yields as evidence of the value of fertilizer. The need for many pounds of plant food to produce a big yield of corn has been overemphasized. In 1962 one of the farm journals showed five farmers who grew over 200 bushels of corn while using comparatively small quantities of fertilizer. I imagine this disturbed some fertilizer salesmen, because the figures did not support the propaganda distributed by sales agencies.

I have grown large yields of corn when smaller amounts of plant food were applied to the soil. The practice did not "wear out the soil," as many people expected. My fertility level (according to tests made by experiment station personnel) increased over a ten-year period of continuous corn.

We don't know how much plant nutrient we must add to produce a big yield. There are so many variables to consider that we can't do much more than initiate plots, apply different amounts of fertilizer, and see what amount gives us the highest yield. We can't take much for granted. Every piece of land is different when it comes to determining the nutritional needs of a given crop.

My methods are not orthodox. They are not based on what I was taught in applied courses in college. They are the result of reasoning about my own experience, reading, and using test plots in the field. My solutions to many problems, when I was able to materially increase yields, resulted from my own interpretations based on whatever fundamental knowledge I was able to gain from papers in the leading scientific journals of various countries. Many fundamental research workers should be credited with having contributed to my thinking. Since I am not writing a reference book, let it be understood that anyone whose thinking seems to agree with mine probably helped me formulate my ideas and deserves credit. I probably have very few original ideas.

I have been accused by people with limited background knowledge of putting out crackpot ideas on soil fertility. Many ideas in this book are at variance with those I was taught in college, and since many present-day workers were taught those same ideas, they naturally are very critical of my interpretations. But since these ideas have enabled me to solve many problems and greatly increase yields on farms over a wide area of the United States, I am glad to assume the responsibility of being unorthodox in my ideas.

I am convinced that I have put together some worthwhile bits of the puzzle of fertility problems, because I did achieve a 100 to 200 per cent increase in yield the first year. Examples of how I have solved these problems make up the major portion of this book.

I have been successful in raising corn yields from 50 to 145 bushels on farms where some experiment station people have failed by their own methods to do more than increase yields by 10 bushels. I feel that I have ample proof in my data and observations, and I can demonstrate the facts with field plots. I am not criticizing all extension teachers. People from some experiment stations are more successful in solving problems than those from others. It depends on how well they have been won over to the philosophy of the fertilizer industry. We have too few people who want to do their own thinking.

We may assume two points of view. One is the fertilizer salesman's point of view: "Increase your yields and field fertility by using more fertilizer." Personally, I wish this were true. I sell fertilizer and I would like to increase my sales by recommending more fertilizer per acre. However, this philosophy ignores the soil and its previous treatment. It is driving people from farms, because the farmer is not making any profit. He gets no response and his costs per acre exceed his cash returns. When a farmer can't pay his bills, it means he isn't growing enough per acre to pay for the fertilizer—so why should he buy fertilizer? This is one method by which fertilizer companies may buy farms. It is the easy way to sell fertilizer. But it is not conducive to building up a sound future business.

I am more sympathetic with the second point of view, with which, I am glad to say, a few agronomists agree: "Test the soil and find out what it needs, then try to initiate check plots to see whether the fertilizer pays off." This point of view increases costs of sales, but it means that the farmer gets enough returns to pay his bills.

This book is written for the layman, particularly the farmer, who has the responsibility of feeding an ever-increasing population. Along with this obligation he has the right to maintain as high a standard of living as any other small businessman. To achieve this standard of living he must grow something to sell, and to do this successfully he must grow more than average yields. As a matter of fact, he must grow as big a yield as his climate permits. The efficiency with which crops are produced will vary among individuals, and will result in variations in farmers' standards of living.

I recently read a statement, made by the head of an agronomy department at one of our universities, to the effect that "any sizable boost in world crop production must be accompanied by a great expansion in the fertilizer industry." I doubt whether we have any proof of this. (Had he said this about the ground limestone industry, I am certain it could have been proven.) I object to such statements because they build up false hopes and befuddle our thinking about the real facts. We certainly cannot back such statements with the facts we have gained from past experience. Such a statement suggests to me that commercial fertilizer should be our main consideration in finding ways and means of furnishing future generations with sufficient food.

In my vocabulary, this is sheer politics. We start off with a hypothesis wrongly derived from existing data, and because we want to go along with a popular notion, we make assumptions which cannot be proven. This sort of thinking has lulled some of our scientists into a smug complacency, a feeling that they have to be right. I feel that much of our past crop research might best be junked and that we should start over with some fundamentally trained, non-political, open-minded personnel—particularly at the administrative levels.

This is a harsh statement, but when I am called a rebel I like to know why. A friend of mine once told me, "Don't worry about what people call you as long as you don't rob a bank. You should be glad they talk about you. The time to worry about what they say is when they stop talking. When that happens, they might as well bury you." I am a rebel because I can't go along with the people who say, "If you want to increase yields 100 per cent, apply twice as much chemical fertilizer." It has been my experience that nothing is further from the truth.

I believe in the use of chemicals to increase the crops grown on an acre of ground. But, since I have been a member of several agricultural experiment stations during the past twenty-five years, I can't get enthusiastic about the things agricultural colleges are teaching. I am sure that they would have trouble proving 90 per cent of the things they teach. But they have done a good job of promoting the sale of commercial fertilizer.

I have always been of the opinion that unless we can show the farmer, by plot comparisons on his own farm, that a practice makes him more money, the practice is of questionable value. Over 50 per cent of the chemical fertilizer used on farms today probably does not return the farmer a penny of profit. This does not discredit the chemicals, but it is a criticism of the people who recommend their use. Too many recommendations are based on hearsay, not on knowledge gleaned from treated plots covering a wide area where many variables exist. Such recommendations may produce a profitable increase on one farm and none on the next.

In dealing with farmers, I have been surprised to find that less than 10 per cent follow college recommendations, while 35 per cent follow what the fertilizer salesman recommends. The remainder are guided by past experience. This last group includes most of the successful farmers. This was very disturbing to me, since I helped establish recommendations when I had the responsibility to do so. I am convinced that the recommendations for the best use of commercial fertilizers might better be classified as propaganda to sell fertilizer. They do not insure that the farmer will make more profit from his efforts.

I lived on a farm until I went to college. I studied what was offered in a four-year course in agriculture. I was not happy with my choice. I realized I should have gone into chemistry, physics, and mathematics for fundamental training, taking fewer subjects labeled "agricultural." I realized that agricultural courses were set up to study the *art* of growing plants and animals, not the *science* of agriculture. Agriculture is only the application of science to soils and plants and animals. Why not study the science first and then the art? That way one could better understand the workings of soil and plants.

The application of scientific knowledge to agriculture is of graduate-study caliber and should be treated as such. We have too many college graduates conducting fertilizer experiments (it would be more accurate to call them tests) who know very little chemistry, but the application of manures and chemical fertilizers to the soil, their effect on the soil and on the growing crop, and their relation to weather conditions, are chemical phenomena which demand understanding of every phase of chemistry—simple inorganic and organic reactions; colloidal, physical, biochemical and complicated organic processes, and complications introduced by fungi, bacteria, soil-inhabiting flora and insects.

The "rebel" label was attached to me because I could not agree with what my professors had taught me. I proceeded to prove that their interpretations of soil and plant workings were faulty. The object of this book is to present my side of the story of how and why I became a rebel in the fields of agricultural practice and the teaching of college students.

I can't agree with common teachings on the use of chemical fertilizer. I have talked with many learned men who have studied in European universities who agree that my interpretations are far more in keeping with chemical law than those of the men who condemn my teachings. And I can prove, by the use of test plots on farms, that a farmer can make more profit with my ideas.

V. A. T.

To

MY WIFE

whose patience and help

made this book possible

CHAPTER 1

Abundant Crop Production and Good Nutrition Must Be Well Integrated

IT IS a safe assumption that there are some things affecting crop yields about which we can do very little. A bushel of corn or other grain crop is made up of starch and storage protein. A bushel of corn (56 pounds) is made up of 6.7 pounds of storage protein, which in turn is carbon, water, nitrogen and many minute quantities of minerals, including phosphorus, sulphur, manganese, magnesium, and calcium. These minerals probably account for less than an ounce of the dry weight of the bushel. Water accounts for 8.4 pounds. Starch, oil and other carbohydrates, including a little fiber, account for 41 pounds. Carbon, taken in as carbon dioxide from the air through the leaves along with some water, is converted by means of sunshine to sugars. These sugars are then converted to proteins.

Clearly, the process of growing a bushel of corn is largely dependent on weather conditions, factors not controlled by the farmer. And although the farmer can't do much about controlling the weather, he can learn how crops respond to the different conditions. (This in itself is a worthy subject for further fundamental research.) Also, to make it possible for the plant to grow most efficiently, the farmer can set the stage by making the soil suitable for the proper growth of the plant. What he does about supplying limestone, plant nutrient materials, and oxygen can be the factor that determines how big a yield he will grow, how good the feeding value is, and how economically he can bring about the process. These are the most complicated problems with which he is faced. He must try to evaluate not only the factors over which he has no control but also the factors that he can control.

After one hundred years of research our average yields are too low to insure a farmer a profit. The level of fertility in soils has a direct relationship to our future food supply; but the importance of the application of commercial fertilizer in assuring our future food supply and its ranking in the evaluation of the many factors that have a bearing on our maximum yields are questionable.

The applications of lime to the different soil types and the use of lime in its various forms are a subject which we have recognized for many years but about which we know very little. We have taken the stand that lime is a necessary evil. But observations I have made and much good supporting evidence suggest that it may be the key to the fertility program of the future.

We have done very little about integrating the factors that have bearing on the yields of crops. As a result, much superficial research of a testing nature conducted during the past seventy-five years is worthless today. Soil fertility problems are being discussed from various angles without regard to soil saturation with nutrient ions, the availability of those ions to the growing plant, the acidity and alkalinity of the soil, drainage, aeration, topography, and general location. Even though we have conducted experiments for many years, we are still at the point where we can say only that results were obtained in such a location under the following conditions (which should be named), and that if we would corroborate such results, we must be sure that the conditions of environment are identical. Unless we do, we will get different yield results. Because weather conditions vary from year to year, it is almost impossible to get the same results with a given fertilizer in succeeding years. We have accepted the need for commercial fertilizers because of "fourscore and ten years" of common usage with no questions asked. How much we must depend on the addition of fertilizer nutrients is something which we still don't know.

Everyone dealing with fertility problems has in mind the possibility of placing the whole fertility program on a balance sheet basis—in other words, know what is required for a crop; test the soil; know the amount of fertility or nutrients available and the amount needed by a plant to produce a given crop; and, by arithmetic, apply the difference from the fertilizer bag. To make this possible, we must start out with one absolutely homogeneous soil and completely understand the chemical changes that take place under variable weather conditions. We must be able to forecast the weather for six months or more. In some areas which depend entirely on irrigation for moisture supply, this would seem to be relatively simple, but there are research problems even where we have moisture under control.

At one time I was involved in a co-operative experiment studying the optimum distance that potatoes should be planted apart. The experiment was conducted for six years. Different locations were used each year on what we thought was homogeneous soil, purposely keeping the plots small to reduce heterogeneity. We found out many things, but we could not tell at what distance apart potatoes should be planted, because every year our results were different. There were too many variables—most of them unknown.

It reminded me of a paper I was once asked to comment on. It was a study on "The Effect of Environment on the Yield of Navy Beans." I don't believe that any scientist, even with all the luck in his favor, could solve this in a lifetime. This man had collected data for three years but had nothing to show for his work. Even to study one factor, such as light intensity (cloudy and bright days), could have taken a lifetime. I don't want to seem pessimistic, but I think we try to bypass the time element by trying to solve everything in one year. So it is with our soil fertilizer problems: we try to set up a balance sheet on soil fertility before we know what the variables are.

There is considerable controversy about the application of

facts gleaned from sand culture experiments to soil-grown crops. We might just as well argue the point between loamy sand and clay soils, clay soils and muck soils, or soils in northern Maine and those in southern Florida. Somewhere along the line we have missed the boat, because we are still a long way from predicting what a crop will do on a given field even though we do have soil tests, ample fertilizer and a good moisture supply.

We must admit that sand culture does at least give us a chance to start from scratch. By at least partially controlling the environment and eliminating soil complications, we can obtain data which we have a good chance of duplicating if we can repeat the same experiment under identical weather (light) conditions. This is almost impossible on a soil. Data to show how much nitrogen, phosphorus and potash is needed to produce a pound of corn under a given set of conditions can be obtained with sand culture. All we need is a large number of pots or containers. By varying each factor in sufficient numbers we can arrive at a figure. We at least have some information which we can put down in the record under one set of light values. If we add one additional variable, such as two per cent of organic matter, we have complicated our experiment.

What we do with that data depends on our knowledge and experience. Perhaps on a sandy soil with no profile differentiation we might be able to duplicate the results in the field, if our moisture supply is adequate. Perhaps we could do this on a soil with slightly more silt in it, or even more clay, or we might even allow for some additional organic matter. Where can we draw the line? If we could run this same experiment on 100 different locations and obtain on 75 of them positive results that agreed with our original data, we could feel that we were making progress. Such data would give us considerable confidence in what we were doing. If we could get 75 out of 100 to show some similarity in a definite response, we would have an exceptionally good batting average; but 50 out of 100 comes closer to present achievement. This gets us down to a 50-50 basis—a guessing basis.

A large number of experiments can take much guessing out of research if there is some similarity in the results. If experiments show big differences in yield, we can have more confidence in our results than if we have to use statistical methods to find out whether we have something or nothing. Some years ago a friend of mine showed me some data on sugar peas. In two out of five years he had gotten a negative correlation, while in the other three years he had gotten a positive correlation. He insisted there was a positive correlation. Statistically, yes; practically, in my opinion, the waste basket was the best place for the data.

To minimize field variability I have assumed that it is wiser to replicate experiments a hundred times to get an average trend rather than depend on one experiment for more exact data. I prefer to conduct the same experiment in 100 locations under the same seasonal weather conditions. If 90 of those fields show the same trend or similar results to a given treatment and only 10 show a variable response, I assume there is a definite response to a given treatment. It seems to me that if, when we are conducting experiments, the magnitude of difference must be determined by statistical methods, we are not treading on very sound ground, particularly when our yields are near average for an area. In other words, I cannot get very enthusiastic about 10-bushel yield increases for an area. It means we have not found the real cause of low yields. If we are going to help agriculture, we must get yield increases sufficient to reduce unit costs materially. In most cases, this means more than doubling existing yields.

It has been my privilege and pleasure to work with outstanding men in the scientific field as well as with many critical, progressive farmers. As a result of my work with people in various parts of the country, I have formulated some definite ideas which in some cases have shaken my faith in past research on crop production.

It seems to me very difficult to establish facts. There are so many variables to contend with that at best our results may only be a good guess. That a man can grow 300 bushels of corn on an acre doesn't prove anything. It simply means that he had the majority of variables in his favor. As one farmer told me, whenever he figured he knew how to grow a 200-bushel corn crop, he would get 75 bushels the next year—with practically the same treatment and rainfall. But I am convinced there are some good reasons why our average crop yields are too low to return a profitable labor income to farmers.

Average crop yields, even though they do not return a profit to the farmer, have in many cases been sufficient to maintain natural surpluses, which in turn have demoralized farm prices. It is my opinion, after many years of experience, that these average yields could be doubled with very little effort. The question arises as to why it has not been done, and whether it should be done. I am only interested in the former. Our economists must grope with the latter. My main concern is to help the farmer to grow a crop so that he has something to sell at a profit. He at least won't starve.

Why do we have such low average yields? Either the farmers have not listened to recommendations from the personnel in their advisory systems or the recommendations have not been sound. It is probably a little of both. As I contact members of our farm population, I realize there are some who don't care; they usually squawk the loudest for government help. There are some who are confused; they hear so many different ideas that they usually do the wrong thing. Then we have the chap who is very conscientious, who believes everything he hears, and when he gets through, his acre costs are so high that even good yields bring him small profits.

A good example is the experience of one grower who cooperated with me for many years. He found that adequate limestone could increase his corn yield by 45 bushels. Subsoiling his fields added another 17 bushels. Minimum tillage added 21 bushels. Plowing his ground and planting corn immediately without preparation when the soil was free of excess moisture added another 13 bushels. Planting more seed per acre added 11 bushels. Using fertilizer solution on the seed added 10 bushels and applying foliage spray one to three times added 10 to 25 bushels. The highest corn yield he harvested was 198 bushels, where five years before he was harvesting 50 to 60 bushels. If he had left out the limestone he probably would still have only 50 to 60 bushels, even though he did all the other things. He said you can't leave out part of the practice if you want big yields.

I visited a man who was bubbling over with excitement because he had a beautiful field of corn with a potential 100 bushels per acre. He told me that he plowed the ground and, without further treatment, planted corn. Then he sprayed with a weed killer. The field was clean. He used 20 pounds of 10–20–10 in solution when he planted his corn and sprayed with 20 pounds of fertilizer solution when the corn was three to four feet tall. He said, "Think of it, I don't have twenty dollars an acre invested and it is the best corn I have on the farm. It doesn't make sense." I told him that it was the type of thing that he would have to do to make money—that he must use his judgment in evaluating what he hears and reads even though the information may come from a reliable source, that he must expect to grow a good crop with fertile soil. Otherwise, why own high-priced land?

Contrast this with the corn farmer who had his experience discussed in a farm paper. He achieved the impossible task of growing 100 bushels of corn per acre. But his fertilizer cost alone was over $70 per acre. After figuring his total costs, his only return was the satisfaction of having grown 100 bushels of corn on an acre of ground. The person who wrote the article forgot to mention anything about costs; but this farmer probably complains that something should be done about the farmer's problem, that there is no money to be made in farming.

I have another farmer who had an idea and decided to prove it. His success should merit the highest praise. He bought a farm in a hilly section of southeastern Ohio which was classed as marginal land. As a matter of fact, he was told that he could not make a living on such land. However, he thought that he knew better. He bought the land and applied large quantities of limestone to the soil before and after plowing, before he planted corn. Several years later I gave a lecture in his community and stated that if a person could not grow 100 bushels of corn without fertilizer, he had better examine his soil for calcium content. I talked with him after the lecture. He told me that he was growing 100 bushels of corn per acre and he still had not used a pound of fertilizer. I told him he probably could have made more profit by applying some fertilizer.

When I realize that actual facts are hard to establish in a field where so many variables exist, I begin to wonder whether we know anything about crop production. To me, the disturbing thing in crop nutrition experimentation is when I take for fertilizer comparison experiment what seems a poor piece of ground and find that the plot I left with no fertilizer yields as well as the one I used fertilizer on. I have had this happen with many different crops.

During the years I spent conducting fertility investigations in the coastal plain soils of New Jersey and Virgina, I found that it was virtually impossible to get worthwhile results from plot experiments unless I first investigated the subsoils to see whether the chemical and physical condition of the subsoil would permit a crop to grow. If the roots could not penetrate the subsoil, there wasn't much use in conducting an experiment on fertilizer comparisons in such a location.

Dr. Jacob Joffee (from the Soils Department in New Jersey) and I conducted an experiment on sweet potatoes in Lakewood sand in the 30's. We found the eight inches below the surface to be very low in calcium, so we applied 1,000 pounds of pulverized limestone per acre in the bottom of the plowed layer as the ground was being plowed. We did this by hand on all but three check plots, which received no limestone. Then we plotted our fertilizer treatments in triplicate over the field. In spite of the fact that we had widely different fertilizer treatments, the only plots that showed any difference were the check plots where we had placed no limestone. Here, the yield was less than 100 bushels compared with yields of 300 bushels—plus or minus 9 bushels—on the other plots. Where we had placed limestone in the bottom of the furrow we had deep root penetration; whereas on the plots where no limestone had been applied we had no roots below the plowed layer. Adding limestone was more important than the kind of fertilizer we used.

DEEP FERTILIZER PLACEMENT

Lakewood sand is a seacoast deposit which has no profile differentiation. These soils are considered very poor for crop production. Because of the openness of these soils it was considered imprudent to apply fertilizer anywhere except in a thin band alongside the plants. Upon careful consideration it was decided to place the fertilizer deep in the soil, 12 inches to 16 inches in the subsoil directly under the row of plants. The reason for this was the fact that nutrients move to the surface during the growing season. Except in the hurricane season, the amount of rainfall is not enough to cause leaching; therefore, the fertilizer left on top never made contact with the roots. It stayed in the surface and caused the roots of plants to stay near the surface, where they soon became short of water, causing the plants to wilt.

When we placed the fertilizer deep in the soil we were surprised to find that the plants did not wilt, even in the hot, dry weather. When we harvested the crop, we found that where the fertilizer was placed deep sweet potatoes produced three times the tubers produced where the fertilizer was applied alongside the plants. Tomatoes yielded so well that the grower took in $3 for every $1 where the plants were side-dressed. These results were obtained where fertilizer was applied. Where no dry fertilizer was applied the yield was still higher than where dry fertilizer was applied as a side-dressing. Where we used one-fifth as much fertilizer but dissolved it in water (100 pounds of 5–10–5 instead of 500 pounds) the crop yielded 50 more bushels of sweet potatoes.

When I came to Ohio, I assumed I had left this type of soil, since I was in the western area, where the soils are derived from limestone. I did not expect to find poor root penetration in the subsoil. But I was mistaken. In August of '56, I saw a corn field which I would have bought at 100 bushels an acre in July. But when the corn was harvested we had 19 bushels of nubbins. I had tested this surface soil and found in it a good level of fertility and adequate available calcium. I checked the subsoil after the crop was harvested and found no roots below the plowed layer. The calcium reading was less than 200 pounds when it should have been 2,800 and the soil was too hard to dig with a shovel. The acidity (pH) was near the neutral point. Apparently, on long-cultivated soils one cannot take much for granted. I brought some of this subsoil into the laboratory for a preliminary test, determined the calcium requirement, and found it needed 6 tons of limestone to satisfy its needs in the surface acre-foot.

I mixed the limestone with half of the soil, filled six coffee cans with this mixture and six with the untreated soil. I planted 5 grains of corn in each. There was no difference in the time of germination, but there were 28 seedling plants in the treated and 16 in the untreated cans. Both lots grew equally well until they were 8 inches tall. Then I noticed some marginal browning on the untreated plants; and when they were 16 inches tall they were yellowish-green. The lower leaves had dried. The plants in the treated soil remained a nice, healthy green up to 24 inches tall before they showed any nutrient deficiencies. The color began to fade, probably from insufficient nitrogen, since I had applied no fertilizer up to this point. Then I applied a weak solution of nutrients to four out of the five cans in each group. Those in the treated soil showed a response in three days; but I got no response in the untreated soil. I was surprised that the plants in the untreated soil made any growth beyond what one would expect from the seed; but when you take such a subsoil, pulverize it, and expose it to the air, you usually get a good effect from the aeration that occurs.

I have had farmers argue with me that they would ruin their soil if they followed my suggestion that they plow deeper to increase the depth of the surface soil. I feel that when a man can't grow over 50 bushels of corn, he is not gambling very much by turning up an inch of this subsoil, unless he is on exceptionally low calcium soil. In this case, the application of 2 tons of finely divided limestone could increase his yield 50 to 100 per cent. This practice could also cause trouble if there was considerable clay and three or four inches of rainfall in a short time. Low-calcium clay can seal the surface so that air is excluded from the roots.

In eastern Virginia, one of the co-operators who followed my advice on deeper plowing and applying limestone, deepened his plowed layer to 14 inches and told me he harvested 80 bushels of soybeans per acre. I did not see the check made on his yield but I did see the beans before they were harvested. They were in 36-inch rows, the plants were between 3 and 4 feet tall, and the pods covered the stems, so I had no reason to doubt his word. He used no fertilizer. I have seen equally good plants grown where the nitrogen was too high, and they produced only 17 bushels per acre. Too much nitrogen has kept many a farmer from growing over 30 bushels of beans to the acre. Last year I checked a field of green peppers which were in flower. The plants were 2 feet tall. I told this grower he had too much nitrogen and that he would be lucky if he picked any peppers. He told me later that he picked no peppers from this field but that he did have a beautiful crop of stems and leaves. He said he never saw such beautiful, large, dark-green plants, characteristic of plants growing on soils where nitrogen is out of balance with the other nutrients.

I traveled throughout the East Coast, from New England and northern New York State to Florida, and evaluated spinach fields. I found tremendous variations in the growth and quality of the crops. Even though one may not know why the differences occur, one does realize that there must be big differences in soil conditions. Weather conditions can usually be ruled out because good and poor fields occur on both sides of the same fences along the whole region. As I inquired about fertilizer practices, I noticed that the best spinach had the least fertilizer and some of the best quality was in soils that had had considerable limestone applied.

While I was at the New Jersey Experiment Station I conducted an experiment in two-gallon crocks. One series was growing spinach on adequately limed soil, while the second series was grown on the same soil with no limestone added. Fertilizer placement treatments were comparable on both series. The available calcium reading on this soil was 400 pounds per acre. The limestone series had sufficient limestone to eventually raise the available calcium to 2,800 pounds. Spinach seed was planted in all the pots and thinned to four plants in each crock. When the spinach plants in the limestone series were 6 inches high, they were harvested, weighed, and then the plants were dried for dry weight yield. There were no differences between the treatments in the limestone series. The effect of the limestone overshadowed every other variable. In the low available calcium series it was quite another story. The plants ranged in size between 2 and 4 inches. The best plants were in those crocks where considerable low analysis fertilizer was thoroughly mixed with the soil. Where the fertilizer was placed in bands to one side of the seed the roots never made enough growth to reach the fertilizer.

While I was with the Virginia Vegetable (Truck) Research Station, a representative from the Beech-Nut Packing Company came to see me about the idea of canning spinach in the Norfolk, Virginia, area. After a two-week survey he decided the quality of the spinach was not good enough to meet their rigid specifications for baby foods.

I went into a huddle with Dr. L. Danielson, our plant physiologist. We decided to work with several growers who grew over 200 acres of spinach.

A survey showed that 1,000 to 2,000 pounds of a complete fertilizer was applied for each crop and two to three crops were grown each year. December and January spinach did not grow freely. We found that the spinach went to seed and sent up a seed stalk prematurely as soon as it was ready to harvest, so that the harvest season was less than a week. This meant that the crop had to be dumped on the market without regard to price.

Dr. Danielson started by taking samples of the foliage and soil and analyzed the foliage. He calculated the amount of nutrients in an average yield, 300 to 400 hampers per acre, and found 187 pounds of nutrients which the plants recovered from the 1,600 pounds of fertilizer that had been applied for the crop. Then, he checked the soil and found a pH of 6.8 but only 400 to 800 pounds of available calcium. The phosphorus and potash readings were very high. All the spinach roots on this Norfolk sandy loam were in the plowed layer.

We decided we needed 2,800 pounds of available calcium on this soil. The grower started to put on limestone at the rate of 2 tons per acre. We also reduced the fertilizer to 500 pounds per acre. In addition we applied 20 pounds of manganese sulphate in anticipation of a deficiency of manganese because of the heavy limestone applications. We found later that this was not necessary because the plants were getting more manganese as we applied more limestone. This, we found later, was because the roots were readily penetrating the plow sole and were feeding in a larger volume of soil.

We could see an improvement in the spinach, but it wasn't until two years later that the farmer grew his big spinach crop.

Before we started I had told him he should grow 1,000 hampers of spinach on an acre. He looked at me, and after what seemed like too long an interval, he said, "Tell me how to do it; I never saw that much spinach." Just before Christmas of the second year we worked with him, he came into my office with a grin on his face. He said, "I just figured out my average yield on the 200 acres of spinach. I sold 220,087 hampers or a good 1,100 hampers per acre. Furthermore, I have a letter from my commission man in New York City. He told me my spinach was the best quality that they had ever handled from this area."

He had applied 7 tons of limestone per acre and was using only 500 pounds of fertilizer for the three crops. He said the pH was just below 7.0.

This situation is a good example of the problem existing in much of the area on the Coastal Plain soils: insufficient available calcium and too much commercial fertilizer. Only by adjusting the available calcium in these cultivated soils can we hope to make some progress in determining the fertilizer requirement. We have conducted too many fertilizer requirement experiments on soils where the controlling factor to crop production was available calcium rather than insufficient fertilizer.

Many of our potentially productive soils are idle. If one travels over the various countries of the world, one cannot help but be impressed by the tremendous amount of land that is not under cultivation. Some of this is too rough to be cultivated, but there is much that is abandoned land, cut up with fences into small areas. Much of it is in old pastures which are not producing enough grass because the available calcium is too low. Much of the land is level and is covered with poverty grass. This land has a high pH but needs 8 or more tons of limestone per acre to make it productive.

Many people who supposedly have the training to be soil and crop specialists catalogue this land as worn-out or too poor to be profitably farmed. I have never considered a soil worn-out just because it is unproductive. There is nothing to wear out. It may be temporarily depleted in fertility, or it may have been mismanaged and permitted to become depleted of available calcium, thus making it unproductive. I feel a soil is unproductive because chemical reactions are not taking place as they should. If we know of what a particular soil consists, we should be able to get our cars back on the track in the proper order, so the locomotive can get them to their destination. This is something that few people seem to be able to do. Too many of our research men get the cart before the horse and then expect to make the horse function properly.

In my early contacts with farm people I was surprised to find that there were people who farmed 160 acres, who, because they were losing money, would go into debt to buy another 100 acres, thinking they would be in a better financial condition. In business (a farmer should be in the same category), volume of business usually is correlated with profits. A big overhead cost divided into a greater number of units of production would result in more profit or make it possible to sell the unit at a lower price. In other words, if you add more machines under the same management, you produce more units of production. This can be true on a farm, if the farmer is a good manager to start with. But if he can't make a profit on 160 acres, the additional land will have to be tilled by hiring more high-priced labor, and he will still be losing money. Here, again, the man's intelligence is highly important. A lazy mentality is sure to lose out.

I have worked with people of various degrees of intelligence. One of these, who I felt was not getting out of his land what he should have with his apparent mental capacity, asked me to work with him. He had read everything he could and was so confused that he couldn't make up his mind what was the right way to handle his soil. Also, he had been misinformed. I told him the first thing was to develop a reasonable philosophy by reading more critically and not believing everything he read or heard. I heard a farmer question an agronomist at a farm meeting as follows: "Doctor, your statement doesn't agree with the article I read in ________ farm paper. Can't we rely on them?" He was answered, "Don't believe anything unless it was written by someone in my department." It has always seemed to me that there is one correct answer to any question regardless of who answers it. So, when the man I was working with asked me why he couldn't grow over 80 bushels of corn per acre, I told him his philosophy was faulty because he read things written by people whose thinking was faulty.

He had been told that he had sufficient lime in his soil and that he should use twice as much fertilizer. I told him that he needed 6 tons of limestone, and that he could cut his fertilizer application in half. When he argued about it, I told him I would be glad to prove it to him by growing a crop with my application alongside his in at least three places on his farm. When we harvested his crops, he had 83 bushels, I had 141 bushels. Then, he wanted to know why he had been given the wrong information. I told him that the good doctor who misinformed him probably didn't know any better. When he got these results three years in succession, he decided to farm fewer acres so he wouldn't have to hire help. And, as he told me later, he was making more money.

We have too many people in our agricultural institutions and in our agricultural advisory systems who are unintentionally mentally lazy. When they get their college degrees, they stop being students. They have certain fixed ideas in their minds, and when they hear or read something that is at variance with their thinking, they not only refuse to examine it critically but go so far as to condemn it as propaganda which someone is peddling to help sell his product.

We have too many people doing research who feel that our salvation lies in the use of large quantities of fertilizer. They are as wrong as is the man who claims we can get along without fertilizer. Most important is the discovery of the factors that control crop production; these can then be dealt with in their proper order. This requires not only clear thinking but good judgment and accurate observation. Plot comparisons should be used freely in making decisions. There is one comparison that should always be made: A comparison between 2, 4, 6, and 8 tons of limestone will always show yield increases, because it has been my experience that at least 90 per cent of the time saturation of 85 per cent of the base exchange capacity of any soil with calcium will materially increase yields.

CHAPTER 2

The Reasons for the Many Effects of Limestone Have Not Been Well Understood

THE USE of liming materials in the production of horticultural and field crops is an ancient practice that dates back to the days of the Roman Empire. Because of a lack of research and failure to recognize its merits in the exchange complex, lime has been used more or less spasmodically as a soil amendment. Its relation to crop production still is not well understood. Lime has been used primarily as a corrective for soil acidity, which probably accounts for the fact that it is still used at irregular intervals and in inadequate amounts. It has never been considered for its calcium content as an integral part of the fertilization program. No one seems to have known that plants' use of calcium and magnesium was only a small part of lime's total effect on the soil. The importance of calcium in the exchange complex and its ability to promote higher yields certainly would have received much more of the attention of American researchers if they had paid more attention to the research done in England, Germany, Holland, and Russia.

There are several reasons why lime has been more or less ignored as a fertilizing material. From the dawn of crop production, land fertility was taken care of by the migrations of tribal herds from worn-out to more fertile grazing areas. These migrations continued for centuries; but we are only now beginning to realize that now we must raise crops on so-called worn-out soils because there are no ungrazed areas to move to. New frontiers of fertile lands are a thing of the past, excepting perhaps those fertile soils in the West which can be brought into production by irrigation. The adaptation of crops to special soil types has only recently been recognized. Some crops were supposed to require special fertilization practices. Soils vary tremendously in their calcium and magnesium content. The discovery of this has given rise to the thinking that certain crops can be grown profitably only in certain areas. Had there been an earlier appreciation of the importance of calcium as a plant food material as well as of its importance as a chemical ion in the base exchange complex, our thinking would have been directed into a different channel. Natural limestone content in the soil has influenced the distribution of the agricultural population in this country. Had we known why this was so, our natural resources of plant food materials could have been preserved by improving the physical conditions of the soil. As it is, we have lost millions of tons of fertility because of sheet erosion and tile drainage. These salts have been deposited in the oceans by streams and rivers; and the process is still going on because of wasteful farming practices.

The introduction of plants intolerant of acid conditions into regions where soils were naturally low in calcium focused attention on the needs of plants for calcium. The introduction of leguminous forage crops as a source of animal feed probably has done much toward changing our ideas about the real function of applying lime to soils.

The use of animal manures as a source of plant food materials for crop plants somewhat alleviated the need for liming materials because much calcium was returned to the soil in manures and crop refuse. We now know that we must add more limestone when we apply manure to the land because of the calcium needed in the additional exchange complex introduced by manure. These manures also contributed to the organic matter content, which tended to prevent calcium from being leached from the surface soil. Wood ashes, land plaster and other soil amendments tended to postpone the day of reckoning for the need for lime.

The substitution of chemical plant food materials for animal manures has focused attention on the fact that calcium replacement had been overlooked. It has had a tremendously stimulating effect on plant nutrient research.

During the past fifty years, experiments set up to determine fertilizer formulas in most cases were carried out irrespective of the lime content of the soil. The close relationship between calcium and high yields is still a new idea. Many fertilizers were formulated for acid soils because many of the soil areas where experimental work was done had their origin in acid rocks. The result was a number of fertilizer formulas which were extremely high in phosphoric acid. It is possible that if these formulas had been set up as a result of experiments on limestone soils, many of our common fertilizers might have carried only half the phosphoric acid content. The fact remains, however, that many agricultural soils are too low in available calcium to produce good yields regardless of the quality or quantity of fertilizer applied. In the past, larger quantities of superphosphate were used, because they contained gypsum, which supplied needed calcium. Potatoes were supposed to be grown on acid soils because scab supposedly followed lime applications. Scab lesions from scab organisms are similar to those formed from fertilizer injury. Much of the scab observed in the past was from fertilizer injury and could have been corrected by heavy applications of limestone (but not hydrated lime).

Of the 1,600,000 acres of land in the United States devoted to only the production of vegetables, less than 500,000 probably have sufficient calcium (lime) to produce maximum yields and high quality. That calcium plays a major role in promoting plant growth has not been widely appreciated and probably accounts for the fact that a million acres of land used for growing vegetable crops are insufficiently supplied with liming materials—in spite of the facts that lime is the lowest-priced material used in crop production and that abundant limestone is within easy access of all growers. But because its value has not been appreciated one of our great natural resources has been insufficiently exploited. Much could be done by the federal government to make available a better quality of ground limestone.

Portable soil acidity testers, from simple litmus paper to the much more complicated indicator solutions and electrical measuring devices, in the hands of farm advisors during the past thirty years have played an important role in promoting better crop growth, but the tests have been misinterpreted so often that we have not taken full advantage of them. These acidity tests have served their purpose for pH determinations, but research has shown that they have not told the entire story. Because of this, many of our soils are very much underlimed. Calcium saturation of the soil and the pH test are not necessarily correlated. On soils that have been heavily fertilized, the pH test may not indicate the deficiency of calcium.

The role calcium plays in reclaiming so-called worn-out soils and abandoned soils which are reverting to the wild state is a fertile field for investigation. Many of these soils are fundamentally productive and, in areas where environmental factors are particularly suitable for crop production, they may be reclaimed with calcium applications in the form of limestone. The experiences of a few investigators indicate that these soils are "worn out" because the calcium level has become too low. The degree of calcium saturation has decreased to the point at which profitable yields no longer can be supported. The physical structure of the soil and the chemical equilibrium between the soil particle and plant roots have prevented extensive root growth.

A large portion of the million or more acres now used only for the production of vegetable crops may be considered depleted soils, in spite of the fact that they are being farmed. Growers are putting up a stiff battle, with the co-operation of government agencies, to make these acres again produce crops of which they once were capable. But too often the main attack involves merely ever-increasing applications of fertilizer. This is the wrong way to maintain a high fertility level year after year—especially when the use of lime instead of fertilizer (it can be done) requires only a nominal expense and is the true solution to our problem. Limestone is an investment that sometimes must be liquidated over a period of years, and many growers are financially situated so that they cannot assume such additional obligations. They are reluctant to initiate the program necessary to return these soils to the profit-producing class, although yield increases the first year will often more than pay for the limestone needed to rejuvenate the soil. One heavy application may last for five to twenty years, if the calcium saturation has been raised to the necessary level.

The advent of the Quick Chemical Soil Test, supplementing the acidity test, has helped to give us better knowledge of chemical soil and plant processes, and growers' experiences have made it possible to evaluate certain practices necessary in bringing worn-out acres back to profitable production. Much information can be obtained by observing farm practices which result in above-average yields. Yet, were it not for the occasional grower who has been willing to study his problems conscientiously and take a chance, methods for improving soils would still be in the testing stage.

One of my Ohio customers brought me two soil samples from a field that had a two-acre clay knob which had never produced anything. He said he had several of these on his farm. He had been told the knob was overlimed because the pH was 7.2. I tested both soils and found the available calcium in the clay knob was 800 pounds. This soil had a requirement of 3,600 pounds in the acre-foot. The other sample, from the lower ground, had 1,200 pounds but needed 2,800 pounds. I recommended 12 tons of limestone for the knob and 6 tons for the lower ground. He applied these amounts and planted corn. He harvested 154 bushels off the knob and 147 from the surrounding area. Since that he has corrected the calcium deficiency on four other areas with equally good results.

Twenty-odd years ago, John Tietje, a neighbor in Marinette County, in northeastern Wisconsin, decided he would try to grow alfalfa, much against the advice of his neighbors. "Alfalfa won't grow here," they all said, "it is too cold." But John had his own ideas. His farm consisted of 3 to 9 feet of sandy, loam soil, covering a dolomite (calcium and magnesium limestone) deposit some 30 or more feet thick. "Lime should not be a problem here," John thought.

The first trial with alfalfa was a failure. The seedlings did not look healthy. Many were killed the first winter. But John was determined to grow alfalfa. He went to a farmers' institute held in the community center house that winter and listened to a specialist from the state university talk on forage crops.

"If you can't grow red clover on your farm, you probably can't grow alfalfa. You may need lime," the audience was told.

"How do you find out you need lime?" a voice from the audience asked.

"Get some blue litmus paper and place a piece between two handfuls of moist soil. If it turns red, it shows that the soil is medium to strongly acid."

I watched John. I knew that he was interested. That noon the group was treated to a dinner prepared by the ladies of the community. The chairman of the meeting took the guests to dinner, and John took a seat across the table from me. It was 20 degrees below zero outside with two feet of snow, but there were sixty people at the tables. There was a great deal of interest in the farmers' institutes in those days. This was a three-day session covering everything relating to dairying and potato farming.

I heard John ask, "Will limestone soil become too acid to grow clover or alfalfa?"

"Yes, it may and very often does at the surface of the plowed layer if you don't use some extra lime."

"Why is that?"

"The soluble or available lime washes or leaches out of the surface and the seedlings don't get their roots into the sweeter soil below until too late in the season, and they don't make enough growth to prevent winter killing. Some seedlings may even die before the roots get started. You will always find a few plants that seem to get a hold."

John told his experience. The specialist said he was pretty sure it was the result of a lack of lime, because this was a sandy, loam soil where lime was apt to leach out of the surface.

The following year John got some litmus paper at the drug store and found that his soil was quite acid. He hauled two loads of by-product lime from a sugar-beet factory fifteen miles away. He had to drive over muddy roads with a team of horses and a "wide-tired" lumber wagon. I can still see those two loads as they went by. He spread the lime with a shovel over four acres of ground, and the next year he got a stand of alfalfa. At the end of the fifth year his field was still producing eight to nine loads of alfalfa.

I watched this experiment with considerable interest because it was something different. We had always depended on manure. If we bought chemical fertilizer, it was a hundred-pound bag for an acre of sugar beets, but to be able to re-use the limestone after once using it to purify sugar in order to grow better crops was something that required serious consideration. We had millions of tons under our farm but it wasn't available. It had to be ground. I appreciated the full importance of the possibilities. This was very revealing—getting results from limestone on limestone soils. When one realizes that millions of tons of calcium carbonate are buried under our soils and that we depend for our food on calcium, one wonders why we were so well provided for.

It has been my privilege to travel over many of the cropping areas of the central, eastern and southern states, conferring with growers. Lime or calcium has been one of the crying needs of most of them. Were it not for the fact that barnyard manure has been available, many more farms would be added to the list of those abandoned because the soil no longer would produce enough crops to pay the taxes. Now we are beginning to realize that the addition of sufficient limestone alone can restore fertility.

If the lime needs of the soils in the New England states and some of the coastal plain states are an indication of the calcium condition of the 1,600,000 acres on which vegetables are grown in this country, the limestone under the farm on which I spent my boyhood days, if applied to these hungry acres in a finely ground form, would produce sufficient profits to pay off the national debt. We would probably be farming fewer acres at a greater profit. And that deposit was only a drop in the bucket when compared with the accessible limestone deposits in practically every state in the union. Limestone seems to be one of our natural resources which has not been overworked and which will never be depleted. It is one of the least expensive materials growers have to buy; but, of those commodities that growers feel they need to grow good crops, it is probably the most difficult to sell.

Some forty years ago I assumed the responsibility of managing a 600-acre lumber company farm in northern Michigan. This farm had been producing timothy hay and pasturage for many years for the logging camp horses which were kept there every summer. However, plans were made to build a dairy and grow all the feed (grain, clover and alfalfa) that was needed to feed forty cows. I soon found that clover and alfalfa were out of the question unless something was put on the soil. The climate was supposed to be too cold. The yields of none of the crops were good.

A county agricultural agent visited us from one of the counties in the lower peninsula of Michigan. He came on a raw, cold, late April day and we sat in the barn and talked about lime and manure and what the condition of the soil probably was. He had good judgment about farming and said that he had been raised on a farm. He was a graduate of the state college at East Lansing. He said he didn't know much about the upper peninsula, however.

"Has anybody grown alfalfa in this county?" I asked.

"I don't know of anybody."

"Do you think it is possible?"

"I imagine so, but you know this farm has been horse pasture so long that it will be hard to get alfalfa started."

"What are the important things to consider?"

"Variety, inoculation, lime, and manure."

"We have enough horse manure to put five good loads on each of the ten acres on that field at the end of the barn. We can get lime from the sugar factory at Menominee, Michigan, and I can send to the university for the bacteria to inoculate the seed, but where can I get the seed?"

"I will get some Grimm seed for you. That is pretty hardy and should stand the winters here. Let's see, you need one hundred and fifty pounds."

"How much lime will I need?"

"Let's go out and test the field."

He took a package of blue litmus paper out of his pocket and handed it to me. He then picked up a sliver from a fence rail and dug a hole, picked up a handful of loose soil, roughly molded it into a ball and broke it in half.

"Put a piece of litmus paper on this half," he said. He placed the two halves together and placed the ball in his pocket. Then he walked down the slope and repeated the operation three more times, each time picking a spot that was representative of certain areas of the field. We walked back to the barn to get out of the cold wind, laid the four balls of soil in a row, and broke each open to expose the blue litmus paper. But the paper was red—bright red in the ball taken from the highest part of the field and a pinkish-red in the one taken in the lowest part of the field. "There you are. That one," he said, pointing to the deep-red strip, "shows a lime need of at least three tons per acre; that one, two; and those two, one ton. You probably need about forty tons of lime. Better have a carload shipped up. Want me to order it for you?"

He certainly was a great help to me. In answer to my question as to why the lower part of the field was sweeter than the high ground, he said, "Leaching and surface runoff by water have brought the lime down the slope." Even in the lower places the lime had leached out of the surface. The subsoil below the plowed depth was much sweeter. The main reason for putting lime on the surface was to get the seedlings started. This was the same story I had heard from my neighbor at home several years earlier.

I got the lime and put it on top of the plowed ground, using a team of horses to pull the wagon, and a large scoop shovel to spread the limestone. The 40 tons of limestone was spread unevenly. The teamsters who were clever at driving teams in the lumber woods during the winter were not so clever with a scoop shovel, and this seemed like a lot of nonsense to them anyway. When the fine dust got in their hair and eyes and through their clothes, it began to irritate their skin. They did considerable cussing, and when they tried to wash the lime off their hands it was like rubbing sandpaper on their skin. I was glad it had been spread. The field was covered pretty thoroughly, except for an irregular strip along the fence. This was fortunate, because it made an experiment out of the field.

The alfalfa seedlings grew well, as did a heavy crop of pig weeds and other weeds common to the locality, which were probably brought there by the manure that we applied. The result was that the field had to be mowed to give the alfalfa seedlings a chance to grow. It wasn't until September that the results of the lime could be seen. The irregular strip along the fence that had no lime was patchy and the alfalfa did not look vigorous. The limed portion of the field was fine. The next year the unlimed area along the edge had no alfalfa. I left the farm that winter and did not see the field until three years later, just as they were about to cut the second crop. It was a beautiful stand, except for the irregular strip along the fence where we had not put any lime. There was no alfalfa there, and the foreman told me that this strip had died out in spots the first winter and was completely dead the following spring. They had taken twenty-six large loads of hay from the first cutting. I saw this field in 1961, some forty years later. The alfalfa is gone, but the crop of corn on the field was very good.

A friend of mine has several dairy farms just outside of Elgin, Illinois, where his tenant is now able to grow tremendous corn crops. Some years ago he asked me to look at his alfalfa field, which he said was very patchy. As we drove over the fields I noticed that the best alfalfa was in areas alongside a gravel road.

Cars traveling along the road raised dust, which was carried onto the field when the wind was in the right direction. The gravel on the road contained limestone pebbles, so that the dust probably had some calcium in it. There was no doubt but that the dust contained something that was good for the alfalfa; but it did not make much difference to the corn crop.

As we drove along the corn field, he said, "Look at that corn, ten foot high, but small ears. It seems like that ground ought to grow alfalfa."

"Yes, it should, except for one thing. Alfalfa needs a lot of calcium. Have you ever used calcium on these farms?"

"Don't recollect as I have. What is it?"

"You know what prepared building lime is? You can get the raw rock, ground fine, but not burned, and apply it to the soil to sweeten it. They call it ground limestone for agricultural purposes."

"Oh, yes!" he said. "Guess I did hear about it. I can get it down toward Joliet."

"If I were you I would apply two or three tons of this to the acre."

In the meantime I took some samples of the soil with me to make sure that I was correct in my diagnosis. The soil tested medium acid, or pH 5.4. The calcium reading was too low. Three years later, on a trip through Elgin, I stopped in to see him again. Almost his first greeting as I stepped from the car was, "How would you like to drive out to the farm?" He had a twinkle in his eye. "I'm the only one who has alfalfa around here. But it isn't quite right yet. Maybe something else is out of line."

As we drove into the yard I saw three large piles of ground limestone in one field. "What have you out there?" I asked with a grin.

"Oh, that is White Gold."

"White Gold?" I said. "Pretty cheap gold, isn't it?"

"Seventy-five cents a ton, if I haul it."

"That seems pretty cheap for ground limestone. How much have you used?"

"Every acre on this farm has had three tons except that field yonder."

We started across a twenty-acre alfalfa field where he had some bad spots.

"See those yellow leaves near those bare spots? Got a few of those scattered over the different fields."

He stopped the car and we got out for a closer inspection. Sheep sorrel was growing where the alfalfa had killed out. Most of the field had a beautiful stand of alfalfa.

"Looks to me as though you haven't used enough lime on these spots."

"But I put on three tons!" he exclaimed.

"This soil is high in organic matter, which means a high capacity to absorb lime (calcium and magnesium). This is a silt loam soil that might take four or five tons of lime to raise the pH from 5 to 6."

"What did you say," he asked—"pH?"

"Oh, yes. That is merely a symbol that is used to designate a certain amount of acid in the soil. By calculation we know it takes a certain amount of lime to sweeten the acid or sour condition from pH 4, which is strongly acid, to pH 5, which is medium acid, to pH 6, which is slightly acid or to pH 7, which is neither acid nor sweet. Above pH 7 we say it is alkaline, like some of the desert soils, which may be pH 9 or 10. We speak of those as alkali soils. That happens where you do not have much rainfall. Now, your soil tested pH 5.4 and I figured it would take three tons of ground limestone to sweeten it to test pH 6.4, in which alfalfa should grow pretty freely."

"You made a good guess except for these spots."

"Yes, it was partly a guess, because if this had been a sandy, loam soil, it might have taken only one ton to sweeten the soil from pH 5.4 to pH 6.4. The amount of lime needed will vary with the type of soil. Let's drive over and look at that limestone. I can't understand why you have these apparently acid spots in the field after using three tons of that limestone."

The limestone proved to be a coarsely ground material. "I can see why you got your White Gold for so little money. I doubt whether fifteen per cent will pass through a sixty-mesh sieve. This material does not act fast enough. If you used a ground limestone of which sixty-five per cent would pass through a one-hundred-mesh sieve, you would not have those bare spots in the field. Of course, it would have cost you more."

"Why doesn't corn need lime like alfalfa?" he asked. "I don't believe my corn is any better where I used lime."

"That is an involved question. It does need limestone, but most of us think it doesn't. The soil needs the lime to make it possible for plants to make efficient use of the fertilizer in this soil. They are thinking about the crop instead of the soil. Many advisers claim that alfalfa is a leguminous plant that is classed with plants that will grow best only where the pH of the soil is between 6.0 and 6.6, while corn is in the medium acid group and will grow at pH 5.5 to 6.0, but not as well as at pH 6.6. This, of course, is not good thinking. Even though corn grows at pH 5.5 to 6.0, it will do much better if grown at pH 6.8, or where the available calcium is adequate because the physical condition of the soil is better. We want to get the soil colloids practically saturated with calcium—as high as 87 per cent, according to some authorities. Alfalfa does not tolerate soluble aluminum and iron in any appreciable amounts, when present in the soil, while corn can tolerate some. Furthermore, alfalfa requires more calcium than corn does."

"What is this aluminum you talked about?"

"It is a funny coincidence, but our best, most fertile soils have quite a lot of aluminum and iron in them. It helps to make them hold their fertility, but iron and aluminum are very toxic to plants if they get into the soil solution on which your plants feed. Iron and aluminum are quite insoluble in the soil—that is, they are not available to plants—if the pH is above 5.5. As the soil becomes more and more acid, below pH 5.5, more and more of this aluminum and iron dissolves in the soil water from which plants must get their food materials. Plants like the potato can grow on these soils because aluminum does not hurt them so much. But for most plants we have to use your White Gold to sweeten the soil, and keep the iron and aluminum where it belongs—out of solution. Grouping plants according to the pH at which they will grow the best is not quite correct. It would be more efficient if they were grouped according to their tolerance to aluminum, because many plants in the high pH group will grow at pH 4 in soils which have practically no aluminum or iron in them. It has been shown that if organic matter is put on a soil too acid to grow carrots or beets, it will make it possible to grow those crops successfully. Organic matter apparently takes aluminum and iron out of solution so that they are not available to be taken in through the roots. Sixteen per cent superphosphate has been shown to have a similar effect on acid soils. That is why our mixed fertilizers are usually high in phosphoric acid. We have to correct the soil physically and chemically."

"I didn't realize I was doing so much to my soil by putting on a little ground limestone."

"I appreciate why you call this limestone White Gold. It certainly would mean a lot more gold to growers if they used more of it. How did you pick the name?"

"My man on the farm said this stuff was like finding a gold mine, but I said it was pretty light-colored for gold. He said he had thought of White Gold."

And so it was White Gold. Gold for crop plants. As I left my friend in Elgin, I pondered this question. What would growers do without limestone, without the calcium that plants take out of the soil to cement their cells together? If it is lacking, the cells fall apart and die. They appear to be rotting. If plants do not get enough calcium, the growing tips die, because there is nothing that will take the place of calcium for this purpose. If calcium is so low in the soil that plant roots can't grow and function properly, deficiencies of other ions may occur.

A pH test is supposed to indicate the calcium level of the soil. However, with the use of large quantities of fertilizer, the calcium requirement is high and a pH test does not give a true picture. In addition to nitrogen, phosphorus, and potash, plants also need large quantities of calcium, but, in addition to being a plant food material, calcium also has many other functions in the soil. Eighty-five per cent of the total base saturation of a soil must be satisfied with calcium. The pH records the strong ions like ammonium and potassium, but not the calcium ions, because they are weak by comparison. Thus the pH test does not give us a true picture.

In warm areas soils naturally have a higher pH, and very often a neutral soil (pH 7.0) may be completely devoid of available calcium. So poverty grass takes over, because it is the only thing that will grow in such soils.

Once I was called upon to consult with a grower of greenhouse cucumbers. The plants seemed to grow fairly well, but had brown roots and water-soaked areas on the leaves, typical calcium deficiency symptoms, but when a pH test of the soil was made, it tested 6.8. Apparently this was not a calcium problem; but the symptoms were there. Furthermore, other crops showed similar symptoms. Chrysanthemum flowers, when cut from plants growing on this soil, wilted quickly and failed to revive when placed in water. The same was true of other plants. Tomatoes did fairly well, but the foliage seemed watery and light-colored. These all were symptoms of calcium deficiency, which could be demonstrated in sand cultures under controlled conditions. A lengthy chemical test of the soil was made. It showed abundant phosphorus and abundant potassium. And according to the soil acidity test the soil had plenty of calcium, since the pH was high. What was the explanation? A calcium test had not been made.

The problem was solved by thoroughly mixing this soil, then dividing it and placing it in a number of ten-inch pots. These different pots were treated with salts which would accentuate or decrease calcium deficiency in the plants. A number of potassium and calcium salts were added to the pots, and cucumber seed was sown in each.

The plants came up normally and showed very little difference until they were about a foot high. Then things began to happen. The plants that received potassium nitrate and the check, along with those which received no additional chemicals, showed slight calcium deficiency symptoms, while potassium sulphate produced very severe symptoms. Plants grown with potassium chloride were free from such symptoms, but the growth was not particularly improved. Calcium sulphate had a slight corrective effect, while calcium chloride, calcium nitrate, hydrated lime, and magnesium hydrated lime completely corrected the deficiency symptoms and produced a particularly well-balanced growth.

Apparently, even though this soil had a liberal amount of calcium, not enough was available to the plants. Furthermore, potassium sulphate accentuated the difficulty, while potassium chloride (muriate) tended to correct it. Calcium chloride also corrected the condition. Apparently some salts which tended to keep calcium in solution were more helpful than those which resulted in insoluble calcium salts. Even though this soil contained tons of calcium, it was not available to the cucumbers. It apparently was not being released into the water solution, where the plant roots could get it.

It has been shown that a high concentration of potassium will prevent calcium from becoming available to plants. When the calcium supplied plants were examined, the root growth was extensive and free from injury on all plants which showed normal growth, while the plants showing injury had poor roots which in many cases were brown or dead. As a result, calcium nitrate, as the source of nitrogen, and limestone were applied to the soils in the greenhouse. The crops no longer showed that soft, watery growth, but made a dark-green, normal type of growth. The interesting thing about this was the fact that before corrective treatments were made on this soil, the plants were very susceptible to mosaic-like diseases which were often mistaken for true mosiac diseases. These all disappeared after the available calcium was increased.

It may be asked, "Why did this soil get into this condition?" The answer seems evident. The grower tested the soil, found the pH was 7.6 to 8.4, and decided it had to be lowered. The quickest method was to use sulphate of ammonia and sulphate of potash, which brought the pH down below 7.0, but which caused such an accumulation of potash in the soil and plants that the small amount of calcium apparently was not absorbed by the plants. The plants took in tremendous quantities of potassium, but not enough calcium to keep the ions in the juices in the plants in equilibrium.

This was further studied in sand culture, where conditions

could be controlled to one factor. The results were identical with what had been observed on the soil. If the plants took in too much potassium or sodium and not enough calcium, the growth was soft and wilted easily. The plants did not build up proteins as rapidly as did those where the calcium was high. Furthermore, these soft plants proved to be very susceptible to certain physiological disorders.

This proved to be a practical demonstration of antagonism, a phenomenon which every student of plant physiology must learn. Antagonism means that if living protoplasm is in contact with calcium and potassium or sodium, at certain concentrations those two materials neutralize each other's toxicity. Calcium or sodium alone would be toxic, but if one part of calcium and ten parts of sodium are mixed, the toxicity of each disappears. This principle, which is demonstrable between calcium and potassium or sodium, also holds for calcium and ammonium or calcium and any other material of a similar alkaline nature, but does not hold between calcium and materials of an acid nature. Thus it is possible to produce calcium deficiency in a soil as well as in sand or water culture, even though appreciable quantities of calcium are present.

Too little attention has been paid to this fundamental principle. In practical terms this means that growers have not used sufficient lime or calcium-carrying fertilizers to maintain a good balance in the soil. More specifically, our fertilization practices have permitted the formation of only partial calcium saturation of the soil complex. There are no ions that will take the place of calcium in this respect. Strontium, under certain conditions, will partially substitute for calcium.

It would seem that the pH and available calcium are not correlated and that it is necessary to depend on a calcium test if chemical fertilizers are to be used intelligently. This is particularly true on vegetable-producing farms where one-half to three tons of chemical fertilizers have been used yearly irrespective of liming practices. This is also particularly true of alkali lands which have a high pH but on which crop plants respond to lime or calcium-carrying fertilizer materials. These soils have so much sodium and potassium that they suppress the action of what little calcium may be present and, as the pH increases above the neutral point, the calcium tends to become less and less available.

There are certain localities where during the years calcium carbonate has accumulated in an appreciable layer at varying depths below the surface. This has not been due to farming practices. Such soils may be low in calcium in the surface layer. Conditions have been favorable for calcium salts to settle out. Also, there may be naturally high calcium soils. Such soils may have an abundance of calcium, but it is necessary to add other fertilizer salts to neutralize the effect of the high calcium. Such soils usually respond to potassium according to the principle of antagonism. They often respond to a 20 to 30 per cent potash fertilizer. They are not very common. The potassium content is high enough to kick calcium out of the exchange complex and make it available to the plants.

Such cases are not as much of an exception as many people think. They occur on soils having a high absorptive capacity, high clay and organic matter content. That is, they can hang onto large quantities of calcium and potassium before any appreciable amount is available in the soil solution for the growing crop, because of the high clay and organic matter content. Peat and muck soils are a good example. These soils release ammonia nitrogen during hot weather and need to release into the soil solution large quantities of calcium. If they are in acid regions where the surrounding upland is quite acid, they may need much more calcium. I have recommended as much as 40 tons per acre and had excellent results.

GREENHOUSE ROSES RESPOND TO HEAVY CALCIUM FEEDING

Much can be learned by growers from experiments to determine their own particular soil needs for calcium or magnesium. A greenhouse rose grower had considerable trouble getting a type of growth that was conducive to a high yield of good-quality roses.

He asked me to examine the plants and diagnose the trouble. He had been a student of mine and had been reluctant to call me in, but he said, "I have been trying to figure it out with the principles which you taught me, but it seems I can't make them fit together so that they do me any good."

I told him, "If we can set up sand culture experiments on the basis of those principles and reproduce certain plant responses, time after time, they must be sound. If you can't do that on this soil, it means there is something about the soil you haven't found out."

"I may have slipped up on my reasoning," he said.

"Well, let's have a look and see what is wrong with your plants." The plants appeared to have too little calcium, so I naturally inquired, "What is the pH of the soil?"

"Practically neutral."

"Have you tested for calcium?"

"No, but I have used calcium nitrate on this bed and it is no different from the others."

This was a little disconcerting because calcium nitrate will correct a low calcium condition. I took my test kit and microscope to his greenhouse and went to work. The plants were soft and watery with very little substance to them. The leaves were small and had a sickly yellowish appearance. I tested for all the things I had equipment for and decided that the symptoms were truly those associated with insufficient calcium, so I asked, "Are you sure those solutions you have for your tester are all right?" He didn't know; so I took a sample of soil from several beds, had the available calcium determined and found the test to show a trace of calcium, low to medium magnesium and phosphoric acid, and high potassium and nitrogen, a typical proportion of plant nutrient materials which, if used in a sand culture, would produce the type of growth found on those roses. What was needed was more calcium, so an application of 2 tons of dolomitic ground limestone was made over all the beds. Then five plots were selected, each fifty square feet, and an additional ton of limestone per acre was applied as a check against overliming. This lime was applied in the middle of August. The prospects of a Christmas crop were poor. Twenty thousand roses was a liberal estimate.

Four weeks later to the day, Frank hailed me on the street and said, "I wish you could see those roses. You wouldn't know the place."

"How are the plots that had the extra ton of limestone?" I asked.

"You know, it's a funny thing, but those plants are growing faster than the others. I gave orders to have another ton of limestone worked in between the plants."

"What are you using for fertilizer?"

"Tankage and horse manure for mulch."

I saw the roses several weeks later and I could hardly believe I was in the same house. The plants had made four to six inches of new growth, the leaves were large and had that rosy-green cast that growers like to see on growing plants. I noticed particularly that "breaks" were evident at the axils of the leaves as well as on the lower part of the stems where those large, heavy canes originate.

Had it not been for these plots, scattered through the house, the grower would not have used more than the 2 tons of limestone. Because these high lime plots continued to grow faster than the others, half-ton applications of ground limestone were made until the houses had received 7 tons of limestone during a year's time.

The first Christmas harvest was 39,000 roses instead of the anticipated 20,000. A year later it was 105,000, and since that time he has had an exceptional yield which has been continuous, regardless of crops pinched for holiday seasons.

The important observation made here was that even though 5 tons of limestone had been added to a soil which was only 6 inches deep in beds underlaid with cinders, the pH remained about 6.6. This soil is considered a heavy, sandy loam of the sassafras series containing considerable organic matter. A quick soil test a year after the first application of lime was made showed a reading of very high calcium (6,000 pounds) and high nitrate nitrogen, low to medium phosphoric acid, and high potash and magnesium. The tankage and manure, used as a mulch, supplied ammonium nitrogen which could be readily absorbed by the roots, because of the near neutral pH and abundant calcium, or could be oxidized to nitric acid, which when neutralized by the limestone was taken in as nitrate nitrogen. These changes, the formation of acids, undoubtedly prevented any tieing up of iron, manganese, and boron, which sometimes is associated with too much hydrated lime in the soil. It did demonstrate the need of applying sufficient limestone to supply the lime requirements on a soil in which heavy mulches are continually maintaining a high level of chemically active organic matter.

One of the dangers of too much lime supposedly is that it ties up the minor elements in the soil so that they are unavailable to the plant. This occurs with burned lime but is not very likely with pulverized limestone. The difference in solubility of the two forms of lime is responsible. I have yet to find a case of overliming injury where a grower has used limestone and some form of ammonia as his source of nitrogen. In almost all cases that have come to my attention when stunting of the plants or definite injury (usually due to some deficiency) occurs following a heavy application of burned lime, nitrate has been used as a source of nitrogen. There is considerable experimental evidence to show that nitrate nitrogen is most efficiently used by plants at a low pH of 4 while ammonium nitrogen is most efficiently used if the growing medium has a near neutral pH. Furthermore, if plants are supplied with only nitrate nitrogen at a near neutral pH, chlorosis due to iron deficiency may develop and may be difficult to correct. Using ammonium nitrogen at these same pH values very seldom gives any indication of iron deficiency. In a soil, ammonium nitrogen does not remain as such for any appreciable period, so that plants probably only absorb a very small percentage of that released in the soil. The formation of nitric acid and subsequent nitrates undoubtedly helps to maintain some available minor elements for the growing plants.

Potash deficiency may occur where too much burned lime has been used, due to the fact that the plants must take in too much calcium for the potash they can get. Experimental results show that some plants grow best when the solution contains five to seven parts of available calcium to one part of potassium. These relationships between nutrient materials in the soil are extremely important to the welfare of the plant, and calcium plays a major role in these relationships.

ASPARAGUS IS A HEAVY FEEDER ON CALCIUM

Crops vary in their calcium needs. Asparagus requires large quantities of calcium. It actually uses more calcium in its growth than it does nitrogen, phosphorus or potash, yet most growers try to grow asparagus without lime. The result is inevitable. Average asparagus yields the country over vary from 80 to 110 crates per acre. Growers who make a practice of keeping the available calcium reading in the soil high or very high are harvesting up to three times that yield.

Two neighbors had an argument as to whether asparagus needed lime. One grower argued that there was enough calcium in their soils, but the other disagreed. Each went his way. I saw the acre yields of these two growers covering three years of harvest. The yields on the limed farm kept increasing each year. The grower who did not keep his calcium reading high had the opposite results. His yields were gradually decreasing.

The highest-yielding bed of asparagus I have ever seen, 350 to 400 crates per acre, had been given two tons of dolomitic ground limestone every year for six years. The pH was neutral. Every year this grower also used manure and mixed fertilizer containing nitrogen from sulphate of ammonia. There were plenty of growers in this area who were using this same amount of fertilizer without lime, and their yields were only average.

Growers should experiment with an acre or even a smaller plot and find out whether lime will give a response. This is as true for all crops as it is for asparagus. It is impossible for any experiment station worker always to make a recommendation that will cover all cases, because he does not know what has gone on before. He can, however, advise the grower about certain trials that might be made to determine what will prove to be good practice for his soil. The question may be asked, "Why put on lime to correct acidity and then add a fertilizer that will make it acid?" That question has been asked by many growers. The answer is the result of experience. See the discussion that follows.

I set up what I considered an ideal experiment a number of years ago on Cape Cod. It involved 6 acres of fortieth-acre plots, each plot replicated six times. The crop was asparagus. The soil was beach sand which at the time had only a few brambles growing on it. We had fertilizer quantities up to 1,200 pounds per acre on different plots. Arthur Brenner, the co-operator, was a good grower and I found him an excellent co-operator. We prepared our plots and set in one-year-old asparagus roots on June 1. We were very careful to have them all of uniform size. The asparagus started to grow uniformly. Less than one per cent of the roots failed to grow.

The last week in August, Arthur called to tell me that something peculiar was happening. We examined the field and noticed that the asparagus on the north half of the field was 2 feet tall and well branched, while the asparagus on the south half had only 6 to 8 inches of growth. The first thing that I thought of was residual fertilizer. When I asked Arthur about that, he informed me that the land had not been farmed for three years.

Before that carrots had been grown on the north half and turnips on the south half. The weeds had been so bad in the carrots that they were plowed under, while the turnip crop was kept clean and harvested. When we checked the plot treatments, there was no difference on either half of the field. The soil test showed nothing. The second year the plants on the south half looked almost as good in late summer as those on the north half after a season's growth, but the fertilizer treatments did not produce any difference in volume of growth that year. I never did hear what happened to these plots in later years because I left the Massachusetts Experiment Station before I could make further observations.

In another experiment that I set up in Massachusetts, I picked a piece of ground between two stone fences. This land had been cultivated for many years. It was a good silt loam with a high organic content, the result of having manure applied with wood shavings for many years. The field was plotted with various amounts of fertilizers and planted with uniformly large asparagus roots. By the end of the first year, I had another failure. The plants along the edges were 3 feet tall, but toward the center of the field they became successively shorter. In the center of the field they were only one foot tall. The surface of the field did not indicate any irregularities. We dug holes, beginning at the center of the field, every 10 feet, working toward the outside. The surface soil, 8 inches deep in the center, gradually increased to 24 inches deep on the margins along the stone fences. We discovered a well-developed plow sole which the roots could not penetrate. The subsoil of this field was low in available calcium. Apparently the organic matter in the surface soil and the depth of the soil counteracted the magnitude of the deficiency of calcium in the subsoil on the growth of the asparagus. The soil composition and amount of top soil proved more important than the fertilizer that we had applied.

I mention these things because most people have the idea that to grow a crop all you have to do is to find out how much fertilizer is needed for good yield, apply the fertilizer, and reap the harvest. Usually the result is a 35- to 65-bushel yield (in the case of corn). Actually, after twenty-five years of research and observation I am of the opinion that chemical fertilizers play a minor role in our yields even though tremendous amounts of fertilizer are applied. The fact is, often the major cost of growing the crop is the cost of the fertilizer. If one puts sufficient thought on the problem, one soon realizes that there are yield-controlling factors which are far more important than fertilizer. Many of the ideas that we accept as proof of the pro-fertilizer philosophy are merely much-repeated idle comments, made in the early days of the industry, which we would find difficult, if not impossible, to prove today.

I was called in on an asparagus problem in eastern Maryland. This gives some idea how much limestone may sometimes be needed. A fertilizer company was being blamed by a grower on a farm on the eastern shore of Maryland for a low yield and poor quality of asparagus. I value this grower's friendship; so I will not give his name.

He had had the reputation of supplying a commission merchant on the New York market with top-quality asparagus. They put up a banner across their booth announcing when his asparagus was available. In more recent years he had lost the name. He blamed his fertilizer company for his troubles, since they had advised him to switch to a neutral fertilizer to avoid applying lime. He was a Cornell graduate and amused himself by reading *Science, Soil Science,* and other scientific journals. He had called in specialists from several experiment stations and had received a different explanation from each as to the cause of the trouble. A representative of his fertilizer company called me in for advice. I immediately recognized his problem as calcium deficiency. This was met by a derisive laugh on his part. He had tested the soil and the pH was okay. It was a little above neutral. I told him it might still be calcium deficiency and that there was only one way to find out: lay out some lime plots and see what happened. This he agreed to do. He took two rows of asparagus across the field and applied one ton of limestone per acre. On the next two he applied 2 tons, on the third pair 3 tons, on the fourth pair 4 tons, and on the fifth pair 5 tons of limestone per acre. This was done at the end of the cutting season around the middle of July.

Late that fall I was invited to inspect these plots. I had an approximate idea where the plots were and as I drove along his lane, which was beside his asparagus field, I was surprised to see two rows of asparagus that had made at least twice as much growth as any others. Since there were ten rows in the experiment and only two showed any improvement over the rest of the field, I did not associate the good growth with the limestone treatments.

When we arrived at the house, the grower asked me whether

I had seen the experiment. I said no, but that I noticed two rows of asparagus far above the others. "That is the experiment," he said. I asked him which treatment showed up. He said it was the 5 tons. When I asked him what he was going to do now, he answered, "I have ordered two hundred and fifty tons of pulverized limestone to put on the field. As soon as I find I can get it, I will start hauling. That means over forty trips to the quarry in Pennsylvania."

Three years later he stopped in to see me on a return trip from the New York market. "Well, I topped the market with my asparagus again this year," he said. It was some eight years later, after I had moved to Virginia, that he stopped by to see me on his way home from Florida. He had spent some months recuperating from some sickness and did not mention asparagus. I asked him whether he was still in the asparagus business. He pointed out the window and asked me how I supposed he had bought that new Cadillac car and who was paying for his sojourn in Florida. "Are you still topping the market?" I inquired.

"Never missed," he answered. "The secret of growing asparagus is to pile on the limestone. I am putting some on every year."

"Do you think that you need it?" I asked.

"I don't know, but I'm not taking any chances. It costs money to gamble."

There is a general feeling among horticulturists that, from a fertility standpoint, asparagus is a hard crop to work with. Results are often confusing. The difficulty seems to be in the starting point. Nothing can be gained from fertility studies if calcium deficiency is the limiting factor. I wasted a lot of time conducting fertilizer experiments on asparagus, until I discovered that asparagus is a lime-loving plant needing a large amount of available calcium because of rapid root growth. Physical and chemical soil conditions, therefore, were controlling factors in its growth. When there was sufficient calcium available, fertilizer treatment showed growth differences which might result in increased yields.

During my teaching days at Rutgers University, I had the pleasure of presenting to a group of short-course students a practical course in plant nutrition. One day during my discussion on lime, I made the statement that a pH test did not necessarily tell the lime needs of the soil, particularly for asparagus. After the lecture, one of the students stopped and said, "Did I understand you to say that a pH test was not enough to determine the lime needed in a soil?"

"That seems to be true from my experience," I answered him. After considerable discussion, he told me his father owned 100 acres of asparagus. They had tested the soil for acidity, and it always tested neutral, so they had never bothered to apply any lime. And if asparagus needed a lot of lime, he wondered whether this yield might not be low because of that. Since he had learned how to run a soil test in his laboratory exercises, I suggested that he bring some samples from the asparagus field and test them for calcium. It would be good experience for him.

He brought twenty samples, tested them, and got no test for available calcium. He couldn't understand it. He brought his father along one Friday morning and we discussed their problem. I suggested that they start applying one ton of pulverized limestone each year until their yields reached a higher level. His father had difficulty agreeing with my ideas, but since they had tried everything else while acre yields continually fell to lower levels, he figured he would not lose anything. I told him not to apply any fertilizer until he could see an increase in an acre strip on their field where he continued to apply the lime. I also told him that he must believe in me because it might take several years before the limestone really began to show results in larger yields. They applied a ton of magnesium limestone every year for seven years. I had heard nothing from him about his asparagus project (although I had passed the time of day with him on many occasions) until he walked into my office five years later with ten tin cans full of soil. I asked him whether he had troubles. He said, "No, but I wanted to check the soil and talk with you." After he told me the following story, I asked him whether I could repeat it. He said, "Do anything you want, since it really is your story.

"My dad had grown asparagus for many years and had always

been able to pay his bills, but during the past five years yields were getting lower each year. When I came to take the short course, Father was in debt to the tune of nine thousand dollars. We had to do something different or stop growing asparagus. When we found out what we should do to increase our yields, we followed your liming program faithfully. The second year after the first limestone application was made, our yield was slightly better and we felt we might be on the right track. The yield has increased each year up to the present time. During the past five years many things have happened. I married my high-school girl friend and my father built me a nice home. We have increased our plantings to two hundred acres and this year, after paying our bills and feeding two families, we had a balance of seven thousand dollars in the bank."

He gave me some figures on their average acre yields, but all I remember about them is the general trend, which I am showing in Figure 1.

FIGURE 1

ASPARAGUS YIELD

One Ton of Limestone Applied Each Year, Starting in 1936

This soil was a Sassafras-Collingston fine sandy loam. Ordinarily it should not require more than two tons of limestone per acre, but as the plants grow they build up organic matter from the decay of the voluminous older roots. This undoubtedly increases the calcium requirement. Also, these soils are acid throughout the depth of the profile, each acre-foot layer requiring as much limestone as the surface layer. Over the years, the lower levels in the profile become calcium saturated, permitting deeper rooting and increasing the calcium requirement even more. I don't suppose the time would come when this grower could feel sure that his soil was completely saturated with calcium, that he had applied sufficient limestone to grow his biggest possible yield. Theoretically, he should have reached a saturation point by 1945 on this soil type, but he apparently was benefiting by additional limestone applications. When he once reaches the saturation point, the yield should level off and continue to maintain that level for ten or more years before it becomes necessary to apply additional limestone.

CALCIUM SERVES MANY FUNCTIONS FOR THE PLANT

If all the functions of calcium in the welfare of plants were enumerated, it would be necessary to start with the effects on the soil and mention, among many others, such factors as overcoming baking of the soil when dry. Limestone improves aeration and drainage and tends to make soils granular. It prevents certain soils from becoming slippery when wet. I was very much impressed when, while I was a student studying soils, the instructor asked us to take two lots of soil and add a pinch of hydrated lime to one and a pinch of soda ash to the other. The two lots were then moistened, packed into balls, and placed on the edge of the furnace to dry. The next class period the two balls were examined. The one with lime crumbled up very easily, the one with soda ash (sodium carbonate) was as hard as a brick and could not be pulverized with the fingers. I believe the soil was picked especially for this purpose, because not all soils would be as suitable for such a demonstration; yet one has only to visit farms to view similar demonstrations on many of them. They say, "I can't get my beet seed up." "Guess I will have to plant radish seed again." "Can't get my seed planted until we get some rain." Those are calcium problems, pure and simple. They may not be easy to correct.

I have in mind an experiment in which a given lot of beet seed was planted on nine lots of soil taken from as many farms, all having different levels of calcium and potassium. The germination on these lots varied from 10 per cent on the soil having the lowest available calcium to 100 per cent on the one having a high reading of available calcium.

Then there is the factor of supplying the calcium needs of the plant. Too many people still consider calcium as a soil tonic and not as a plant food material. They still think of calcium as being a corrector of acidity and determine the calcium level by the pH test. Plants are still better indicators of the available calcium level than laboratory apparatus, and if the responses of plants do not seem to fit in with our theories, perhaps we had better overhaul our theories.

Calcium, when once taken into the roots of the plants, goes to work. If there are acids present it ties up with these. Plants like tomato, spinach, and asparagus, which have oxalic acid formed in the protoplasm, contain calcium oxalate crystals, which can be seen with a microscope. Were it not for this function of calcium, these acids would soon kill the plants. Then, too, calcium has an effect on the proteins of the plant cell, keeping them more or less stabilized. It tends to keep the proteins properly suspended in the cell sap, while potassium, sodium, and ammonium tend to keep them highly watered or hydrated. This is probably one of the reasons why it is necessary to maintain a certain balance between calcium and other nutrient materials. Protoplasm with too much calcium or too much potash probably would not support our plant processes very efficiently. Too much calcium tends to dry out or harden plants, too much potassium tends to soften them excessively. This is probably an indication of why fertilizers carrying 20 to 30 per cent potassium are needed on soils having high available calcium, especially for those plants which have a high potash requirement. This is difficult to understand and to prove, because under variable weather conditions it is difficult to get an accurate reading on the potassium content.

Calcium also combines with pectic acid to form the cementing material which holds the cells together. None of the other materials which are absorbed by plants could function in this capacity, because they would not form insoluble compounds. Calcium must be continuously available because plants must have a steady supply. It does not move around very much in the plant.

When all the functions of calcium are grouped together, the end result is the manufacture of protein and sugar in the plant—and food for humanity. Experiments show that if calcium is not present in sufficient quantity these processes are interfered with, and the amount of sugar starch and protein formed is materially reduced. Because of the importance of calcium in human diets, the amount of calcium which can be taken into the plant and stored involves a major consideration. It is possible to produce plant food products that have a minimum amount of calcium on soils that are too low in available calcium. Such crops are not profitable for the grower nor do they satisfy the requirements of a good food.

There is a general opinion that horses raised on Kentucky bluegrass are well nurtured. Kentucky bluegrass is a high-calcium grass. Perhaps there is a thought worthy of serious consideration. I saw a carload of 300-pound Hereford steers unloaded in eastern Virginia and placed on a well-limed pasture. Seven of those steers were placed on unlimed pasture. Both lots had plenty of feed and water. Three months later I saw those same steers. Those on the limed soil were fat and slick. Those on the unlimed pasture had grown larger but looked as rough and scrawny as the day they were put on this pasture.

My friend in Elgin calls it White Gold. I wonder whether even that signifies the value of the millions of tons of limestone available for better food plants for human and animal consumption.

THE IMPORTANCE OF CALCIUM FOR LARGE YIELDS

To talk intelligently about soil fertility and crop yields, we must understand about soil and plant colloids and base exchange phenomena. Nutrient ions necessary for plant growth must be in solution so that they can be absorbed into the roots. Soil and plant colloids help to store these nutrients in the soil and in the plant. They make possible the base exchange phenomena, which makes it possible to apply large quantities of lime and fertilizer to a soil, which can then hold it in readiness for the plant when it needs it.

When we apply limestone and mix it with the soil, we have a mixture which is only partially ready to support a good crop. Not until the calcium and magnesium in the limestone have disintegrated and become part of the colloidal complex in the soil through base exchange reactions does the growing crop benefit from the calcium and magnesium in the limestone. If limestone is applied to the soil and the ground remains dry, the limestone remains ineffective. If the limestone is too coarse, it may not be effective very rapidly.

Good plants and crops can be grown in pure sand. A sand culture is nothing more than coarse sand to which a weak nutrient solution is added. The plants are actually growing in damp sand, but it is necessary to apply nutrients every day because there is no colloidal material (clay, organic matter) to prevent the salts from burning the roots or building up a high specific gravity.

As soon as a little chemically active colloidal material in the form of very fine clay or organic material, like milk casein, is added, we no longer have a sand culture—we have the beginning of a loamy sand which can soon become a sandy and even a silt or clay loam. This adds complications to our culture. We have introduced materials which make soil acidity and base exchange phenomena our controlling factors. It is necessary that we know the nature of these colloids.

Colloids in the soil are both mineral and organic. Mineral colloids consist of mixtures of iron and aluminum oxides with silicon dioxide, which remain stable (remain out of solution) above a pH of 4.7. They may be a continuous jellylike film or they may be large structural molecules. Organic colloids consist of carbon compounds, usually combinations of proteins and amino acids in combination with humic acids, the last stable products in the decomposition of organic matter. They, too, may exist as jellylike films or as particles of large, complex molecules.

The important thing is that these colloids are surrounded by millions of negative ions or charges, which in the natural state are in balance with hydrogen, a positive ion. Hydrogen is a very weak ion and is readily replaced by any other basic ion, such as calcium, magnesium, potassium, sodium, and ammonium.

The number of hydrogen ions so attached in a given volume of soil, along with those existing as free acids or alkalies in the soil, determines the pH of the soil. In order to be able to give a usable figure we use the logarithm of the total number of hydrogen or alkaline ions. A soil with a pH 4.7 has millions of hydrogen ions or charges. This soil would be very highly unsaturated. Any of the basic materials could be applied to saturate this soil. Soda ash could be applied in sufficient quantity to sweeten it and the soil would be sodium saturated. It would have a pH of 7.0 but would not grow a crop. (See the articles by Gedroiz.) Such a soil, if kept dry, would make good bricks. We have formed such soils in the past by the use of excessive amounts of nitrate of soda on sandy loam soils. The same thing could be done with anhydrous ammonia, caustic potash, or the oxides of the other basic ions. However, the only one that could be used if good crops are to be grown would be calcium—such as in some of our early limestone soils. Scientists who have studied this problem (see articles by Gans, Gedroiz, Hissink, and Way) say that at least 85 per cent of the basic material needed on a given soil must be saturated with calcium. For some reason the calcium ion, of which we have unlimited quantities stored on this earth, has properties that are especially suited to support life, whether it is plant or animal life. Since there is such a preponderance of calcium, we might assume that all life evolved on this earth adapted its functions to calcium rather than to other minerals.

Thomas J. Way, an English scientist who worked in the early 1850's, R. Gans, a German scientist in 1905, D. J. Hissink, a Dutch scientist in the 1920's, and K. K. Gedroiz, a Russian scientist in the 1920's, are responsible for our fundamental information on the theory of why calcium is important in soils and why different soils need different amounts of calcium to make available to the growing plant the calcium that it needs to ensure maximum yield under varying weather conditions.

Since the stage was set on which to build a profitable crop program, such men as Wheeler, Bert Hartwell, E. Truog, Sante Mattson, Jacob Joffee, W. P. Kelley, Hans Jenny, Michael Peech, Marshall, and others have contributed to our understanding of how this calcium ion functions in the soil. As a result of my graduate studies with Dr. Sante Mattson, I feel very strongly that there is a parallelism between the relation of calcium to the colloids in the soil and the relation of calcium to the colloids in the growing plant. Plants which have their colloids saturated with calcium apparently make better food for animals.

The process of substituting calcium, through the application of limestone, to replace the hydrogen on the soil colloids is referred to as base exchange. The base exchange complex in connection with the soil colloids is responsible for all our fertilizer problems. The greater the quantity of base exchange material that exists in the soil, the more complex the soil becomes.

The application of nitrate of soda, muriate of potash, or anhydrous ammonia to a soil immediately sets up a chain reaction whereby calcium is released into the soil solution, making it available to a growing plant. Any one of these materials can replace calcium in the base exchange complex, and it is not known whether the beneficial results are obtained from the replaced calcium or the material supplied to the soil.

Muriate of potash, a common fertilizer ingredient, has been involved in our fertilizer experiments because it has so many possible effects. I witnessed an experiment in which numerous plots were treated with various amounts of limestone—from 400 pounds to 6 tons—in increments of 800 pounds. Then, one half of each of the plots was given 300 pounds of muriate of potash. Alfalfa was grown on the plots.

The potash doubled the yield on the low-limed plots, but the total yield was less than a ton per acre The intermediate-limed plots did not show as much increase attributable to the potash; and where heavy limestone was applied there was no increase in favor of the potash, even though the total yield was six tons of hay. In other words, calcium was the controlling factor, and the value of the potash in the low-limed plots was to kick out of the base exchange complex the calcium which the alfalfa needed and absorbed from the solution.

The application of calcium-carrying materials saturates the base exchange complex of the soil and becomes the keystone to efficient crop production. In it lies the secret of our future food supply.

Base exchange of the soil means nothing to the farmer. And yet, everything he does to his soil affects it. Even though colloids are a very minute part of the soil, they control crops more than anything else. If we talk about the pH of the soil, we are primarily concerned with the ratio of negative (−) and positive (+) charges on the chemically active colloids in the soil. If the charges are equal in number, we have a neutral soil in the true sense; but practically, the pH test may be below or above the neutral point. The minus and plus charges may not be the same strength, which would influence the pH test toward the acid or alkaline side.

The most minute part of a dry soil is the colloidal matter. Milk is a good example of a colloidal solution. The curd in fresh milk is a colloid, and as long as it is sweet, the curd remains suspended in water. When milk sours, the colloids (curd), or casein, settles out. It is a protein colloid and it has negative and positive charges on it.

Organic matter (crop refuse and manure) in the soil oxidizes and ferments to release proteins and amino acids along with other products and actual acids from fibrous and starchy material liberated from the organic complex. These all mix with the mineral (colloidal clay) to form a very complex compound, which in some cases actually combines to form a complex mineral protein taking on millions of negative charges. The decomposition of organic matter, crop refuse, and manure results in finely divided molecular compounds, often referred to as the *humus* in the soil. When it is dry, it is a very fine dust. When wet, it may become a colloidal solution in water.

If we were to take all the ammonium, potassium, sodium, calcium, magnesium, manganese, and other basic elements out of the soil, we would have a 100 per cent unsaturated base exchange compound surrounding the fine sand particles; and the pH in a temperate climate would be close to 4.7, if it was all mineral, or 6.8 in southern soils. If it had a lot of protein mixed with it, it could have a lower or higher pH when completely unsaturated.

When the base exchange complex is unsaturated, it is not possible to grow a crop on the soil. Too much iron and aluminum would come into solution, making them available to the plants by making them soluble in water and, since they are toxic, they would poison the plants. That is why limestone is applied. The limestone must be very finely ground so that it will come into contact with as many soil particles or molecules of base exchange substance as possible. That brings about quick interaction with the limestone.

The calcium and magnesium from the limestone exchange partners with the hydrogen on the soil colloid, making a new compound, a calcium-saturated base exchange complex having a neutral pH, practically 85 per cent saturated with calcium. If there is some magnesium, then the magnesium accounts for some of the saturation in place of the hydrogen, but for the best yields it should not be over 10 per cent at the most. A very high magnesium limestone will cause more magnesium to be held on the colloid.

This complete saturation immediately starts oxidation in the soil, and the minerals begin to disintegrate to release potassium, phosphorus, manganese, and other elements. Some of these then exchange places with the hydrogen or even calcium on the colloids and, through the base exchange phenomena, make calcium available. Thus, calcium comes into the soil solution, along with the mineral constituents needed to make up a balanced nutrient solution. If practically all the hydrogen is replaced with calcium, we have changed a worthless soil to a highly fertile, productive one—without applying fertilizer.

The displaced hydrogen has joined with the carbonate to form a carbon dioxide–water mixture. The carbon dioxide then may be released into the air, or it may work on the limestone or other minerals, as a very weak acid, to carry on the weathering process which releases plant nutrients.

Water, sunshine, and temperature can make such a soil produce large crops. When certain types of dry fertilizer or fertilizer solutions are applied to unsaturated (high hydrogen) soils, absorption of the basic elements into the colloids prevents plants from absorbing them, and nutrient deficiencies occur. Acidic ions or negative ions such as chlorides, sulphates, nitrates, and phosphates are left in the soil solution, but since they outnumber the basic materials, we find a solution badly out of balance. Some plants can make some growth on such soils. It might be possible to grow 20 bushels of corn, 5 to 7 bushels of soybeans, 10 bushels of wheat, 75 bushels of potatoes. But when we apply 5 to 20 tons of limestone on such a soil and mix it in, it immediately boosts those yields four to five times, because the soil solution has been brought into a balance better for crop production.

This improved balance also means that we have improved the environment for bacteria, so that more nitrogen is manufactured from organic matter. We have speeded up chemical processes, so that oxidation of minerals is speeded up. This releases potash and phosphorus, along with other elements needed by plants.

If soils are adequately limed, nothing should be wrong with them, except possible element deficiencies needed for specific growing crops. But there are other things that affect soils which can prevent us from growing high crop yields. They are all indirectly associated with partially unsaturated soil colloids.

According to physical-chemical laws, there is a water-ion relationship which affects soils and crops. Each ion has an affinity for a certain number of ions or molecules of water. Calcium has a small number, potassium has more, sodium still more, and lithium still more. Others fall somewhere in between. Colloidal clays and protein will swell up in varying degrees, depending on the ions hooked onto them. A calcium-saturated clay has a low degree of swelling because calcium has only three molecules of water. The same amount of clay saturated with potassium will swell more, because it has five molecules of water; sodium clay swells still more, because it has seven molecules of water. Such soils, when dry, will crack, an automatic aid to air penetration and some oxidation. Drying out a soil has a temporary mellowing effect, just as freezing does. The effect of fall plowing helps to mellow cloddy soils, because it helps to freeze those soils. Salt marshes have extremely wide cracks when they dry out, because the colloidal matter is so heavily saturated with sodium or hydrogen. They also bake very hard. Brick manufacturers have found that the hardest bricks can be made by mixing soda ash with the clay. A clay saturated with calcium makes soft bricks and will crumble. Thus, a clay soil heavily limed becomes crumbly and won't become hard when it dries.

This same quality in the soil affects the quality of the crops. The same chemical reactions exist between proteins and ions in the tissue of the plant as between the soil colloids and the ions in the soil solution. Thus, corn will grow on a soil well saturated with calcium, when the plants can absorb calcium freely. The calcium is held in the cells by the proteins and a high proportion of proteins to amino acids is present. The maximum amount of dry matter per 100 grams of plant tissue is produced when adequate calcium is present. If the plant can't get enough calcium, it absorbs more potassium, sodium, ammonium, magnesium, or other ions. These all have more molecules of water hooked to them than when calcium was present, so the plant growth becomes more lush and has less dry matter per 100 grams of green material. Under these conditions, the ratio of storage proteins to amino acids is lower and the feeding value of the crop is lower. This often happens when growers use too much nitrogen.

Cattle feeders have told me that corn grown on well-limed

soils will produce more beef or milk per pound of silage than corn grown on soil that does not have sufficient lime in it. It has also been shown that corn grown on well-limed soil will not get moldy and has less shrinkage in the crib than corn grown on inadequately limed soils. This is all associated with the "bound water" effect, or we may even say the colloidal base exchange phenomena which exist in the plant. Corn grown with too little calcium won't mature as quickly, is slow to dry out, and readily absorbs water in a damp environment after it has been dried to 15.5 per cent moisture.

From this it may seem that soil and plant colloids (clay and proteins) practically control a farmer's fortunes. They have a direct bearing on his net profits, and the condition of these colloids with respect to calcium pretty much controls his health and that of his animals. Therefore, the proper saturation of the base exchange complex, whether it is in the soil or in the plant, is the keystone to crop production.

The interacting forces established or existent at any given time in the quality of base exchange, or "buffer system," as many refer to it, determines how readily the plant can get its necessary plant food materials out of the soil. Thus, it controls the yields of our agricultural crops. With good farming conditions, oxidation in the soil, and adequate moisture, any cultivatable soil may grow 300 bushels of corn without supplying any appreciable amount of fertilizer.

Criticism against the use of dry fertilizer does not condemn it; it is a criticism against how it is used. Fertilizer cannot be used to grow crops when the plants are grown in pure white sand, because there is too little calcium present. In soils calcium is just as important. Indeed, it is more important than fertilizer in setting the stage for big yields.

Calcium Comes in Many Forms. Liming materials include any calcium material that lowers acidity when applied to the soil. The following may be classed as liming materials:

1. Limestone—carbonates (40 per cent calcium in purest form).
   Marble dust and chalk—high in calcium.

Calcite—high in calcium.
Dolomite—contains a high proportion of magnesium.
Oyster shell—high in calcium.

The effectiveness of these depends on how finely they are ground. If they pass through a 100-mesh sieve, they are good for soil application. They may be coarsely ground, which makes them slowly available, or pulverized, to make them almost as active as hydrated lime.

2. Burned lime—oxides. Very active (70 per cent calcium in purest form).

Each of those under No. 1 may be burned to give the oxides of calcium and magnesium. All can be bought in various parts of the country.

3. Hydrated forms—hydrates. Very active (54 per cent calcium in purest form).

Each of those under No. 2 may be air slaked to make the hydrated forms. These contain more calcium per ton than those under No. 1.

4. Special forms.
   Shell marls—carbonates. Soft and low grades may be slow-acting.
   Lake marls—carbonates. Gritty and slower-acting.
   Slag—basic—oxides. Active.
   Slag—Thomas—calcium and magnesium silicates. Moderately active.
   Wood ashes—oxides. Active.

Most important, from the standpoint of soil fertility, is whether they contain both calcium and magnesium.

We must replace the hydrogen on the colloids with calcium and magnesium. This means attaching calcium ions and a few magnesium ions to the mineral and organic colloids (clay and organic matter). Limestone also neutralizes the acids formed by oxidation of organic materials, sulphur, phosphorus, ammonia, and other ions.

To understand lime and its effect on the soil, we must appreciate what the soil consists of.

A sandy loam soil contains rocks, pebbles, gravel from coarse to fine; sand of varying degrees of fineness, much of it minerals;

silt of varying degrees of fineness and mellowness; clay; some minerals and some quartz of varying degrees of fineness, some of which is colloidal or chemically active; organic matter—roughage decomposed by bacteria to become fine enough to be colloidal; minerals in sand, silt, and clay including calcium, magnesium, iron, aluminum, potassium, ammonium, nitrates, phosphates, sulphur, boron, manganese; and soil organisms, bacteria, and fungi.

We have two general soils. Limestone soils come from the calcium and magnesium carbonates, and acid rock soils have their origin in sandstone, shale, and granitic rocks. The composition of the two general types may be something like this. There are many gradations because of glaciation.

	LIMESTONE SOIL	ACID ROCK SOIL	
	Sandy Loam	*Sandy Loam*	*Clay Loam*
Sand	50 to 80%	50 to 80%	20 to 50%
Silt	10 to 20	10 to 20	20 to 40
Clay	1 to 10	1 to 10	30 to 40
Humus	0.5 to 2	0.5 to 2	4 to 6
Calcium carbonate	11 to 14 (110 to 140 tons)	None	None
Magnesium carbonate	0.5 to 8	None	None

The big difference between them is the amount of *limestone*. That is the reason why limestone soils are potentially more fertile. It takes less limestone to make them productive. They start with their colloids partially saturated with calcium.

We are primarily interested in the clay, humus, and salts, because their relative condition affects the growth of the plant. Much of the salt is usually in solution in a moist soil. Colloidal humus and clay are not soluble but remain in suspension—just as the curd in milk stays in suspension—and are active chemically. They respond according to the laws of colloidal chemistry.

If you place a handful of soil in a glass of water and stir it up, the last material to settle out is the colloidal material. It may stay in suspension for several days. Imagine clay and humus as being a series of shelves made of iron and aluminum, and the stuff on the shelves to be the ions such as calcium, magnesium, potassium, sodium, manganese, and so on. The shelves are deep and the ions on the front may be obtained more readily by the roots than those on the back. Now, imagine the root of a plant being a truck that backs up to the shelf to load up. It needs certain ions. If it gets what it needs freely, the plant grows normally. But suppose those shelves are loaded with potassium and nothing else. Then the plant doesn't get calcium and magnesium. It gets too much potassium and stops growing. But suppose the shelves are almost empty and only hydrogen ions are present. They are gaseous, and the plant can't grow by taking in gas. In addition, the bench begins to deteriorate and the root takes in parts of the shelf—iron and aluminum. The root shrivels and dies. It is poisoned. In other words, we must keep those shelves strong enough and full of calcium, magnesium, and potassium—in the right proportions. If the minerals in the soil don't supply those ions that keep those shelves filled, we must add them in the form of fertilizer. Calcium is the one most often lacking. We have to put on limestone to supply the calcium and magnesium.

How does lime affect the physical condition of the soil?

An acid soil low in calcium does not permit water to drain away. When it is wet it becomes smeary. When it dries out it becomes cloddy. A high pH may be brought about by sodium, potassium, or ammonium, whether there is calcium present or not. At a high pH, such a soil is slippery when wet and bakes hard when dry. The colloidal jelly holds too much water.

A soil sweetened with lime is not smeary when wet and it does not bake hard when dry. It holds only a small part of the water because the calcium ion does not attract water. A soil low in calcium dries out slowly. A soil high in calcium dries out quickly. Thus, a well-limed soil is much better aerated.

A soil may be sweetened temporarily with certain fertilizer ingredients, such as sodium, potassium, or ammonium. In such cases we get a phony pH which sweetens the soil but may cause certain deficiencies—such as calcium and magnesium deficiencies.

According to published experimental data from research in

England, Germany, Russia, and New York and Florida, it is necessary that a soil be limed to a pH equal to 85 per cent of its calcium requirement to support best conditions for growth of crops. For instance, if a sandy soil has a calcium requirement equal to one ton of limestone in an acre-foot, it is necessary that 1,700 pounds of limestone be added to bring the top 7-inch layer into good condition. And to improve the soil down to a depth of 3 feet, we would have to use approximately 7,600 pounds of lime. The limestone that is necessary to bring up to 85 per cent a clay loam having a calcium requirement of 4 tons of limestone in the upper 7 inches can easily be figured. In an acid soil it may be necessary to put on 15 tons of limestone per acre to supply the necessary calcium to a 3-foot depth. Maximum growth may not be obtained unless this is done.

The purpose of deep liming is to encourage deep rooting. When root growth is compared to a naturally acid limestone soil, the importance of a thorough liming program is realized. The following things happen when adequate limestone is applied:

1. Ready penetration of water by dehydrating the exchange complex
2. Good aeration and oxidation (which goes along with calcium saturation)
3. Opportunity for up and down movement of water in the soil, resulting in better aeration and greater workability
4. Extensive feeding area for the plants
5. Opportunity for the roots to reach a water table in dry weather

PROPER SATURATION OF THE SOIL COMPLEX WITH CALCIUM CONTROLS YIELDS

"Saturation of the soil complex" is so much Greek to most farmers. If you fill a ten-gallon pail full of sand, the space in the pail is saturated with sand. If limewater is added to the sand, it fills the spaces between the sand grains and the sand is saturated with limewater. If you add organic matter, acid, clay, and silt to the sand until the spaces are filled, you would have the sand saturated with organic matter, clay, and silt. Then we would have something resembling an acid soil. The organic matter and colloidal clay still contain minute cavities which are lined with millions of negative charges, each one holding a hydrogen ion. We have a situation similar to a run-down battery, which is useless. The soil also is useless.

If we make up a water solution of calcium chloride, magnesium chloride, and potassium chloride, it will be neutral (pH 7). If we pour this solution over the soil and leave it for an hour, then drain the water off and test the soil, we will find it to be neutral. The calcium, magnesium, and potassium will have replaced the hydrogen ions, making the soil neutral. In the field we do this by adding limestone and using the help of natural rainfall to wash the limestone into the soil.

In other wòrds, before we poured on this acid mixture, we had unsaturated or acid clay and organic matter. After we added the limewater the soil became sweet, because we saturated the clay and organic matter with calcium, potassium, and magnesium. There were still empty spaces where water and air could be held. Before we added the limewater, the sand probably would not have permitted seed to germinate, but with limewater the seed germinated and supported the normal growth of the seedling. Like a well-charged battery, it is ready to go to work—just as a fertile soil is ready to grow crops.

On soil that has not been farmed, we can determine the lime needed to saturate the organic matter and clay (referred to as the soil complex) by determining the acidity. If we know how much active organic matter and how much clay we have in a particular soil, we can calculate how much limestone to add in order to saturate the soil complex. A sandy soil usually has very little organic matter or clay; therefore, the amount of limestone needed would be low. As the clay and organic colloidal complex increases, the lime requirement increases. Thus, a clay-loam soil might require 10 tons of limestone whereas a sandy soil might require only one ton for a change of pH from 5 to 6.

Potassium, sodium, and ammonium salts can all influence the acidity test. They are more active than the calcium ion. If a soil is saturated with sodium, the pH might be neutral but the soil would still need a heavy application of limestone, because calcium saturation determines the growth of plants.

A strongly acid soil has a great number of negative electrical charges, and the sum total of the charge or pull in the soil could be so great that positive ions such as calcium, magnesium, or potassium might actually be pulled out of the cells in the roots, thus preventing the plants from growing. Plant growers have noticed that plants growing freely in a good soil will be very slow to start growing when placed in a poor soil. I have been told that if you apply nutrient ions to the foliage of plants growing on poor soils, in a matter of minutes those ions may be traced through the stems and roots and out into the soil solution. We must think of this soil-plant relationship as one tending to set up a neutral balance between the base exchange mechanism in the soil and the isoelectric mechanism formed by the proteins in the plant cells. As long as we have equal numbers of negatively and positively charged particles, we have a neutral balance in which only an exchange between positive ions might take place—as would be the case if there were an abundance of calcium ions in the soil trading places with an abundance of potassium ions in the roots. In other words, there is a continual movement of ions back and forth until perfect equilibrium or balance is established. Since the plant is growing and establishing more proteins with more new charges, the possibility of a true equilibrium cannot occur until all growth ceases and no more new charges can be formed. Thus, we can think of the growth of a plant as the result of ions being transferred from a saturated to an unsaturated condition. From this, we can see that if the growing crop is to have enough of any one ion, like calcium, the soil must first be heavily charged with calcium so that it will be readily available to be transferred to the waiting charge on the protein in the cell of the feeding root. The ease with which this equilibrium can be maintained could account for the many problems we see in fields where seed doesn't germinate or satisfactory growth of the seedling is not made if the seed does germinate.

It is possible to grow plants in pure sand or in glass beads, but here we must maintain a nutrient level comparable to a solution having a pressure of less than two atmospheres, so that the necessary ions will move from the water bathing the roots to the proteins in the protoplasm of the root cells. This is far different from a soil having much chemically active clay and organic matter, including proteins, which are tenaciously holding the ions by a force of varying magnitudes. This force becomes weaker as the saturation point is reached. When it becomes weak enough to equal that force exerted by the protoplasm in the root cell, then an exchange of ions occurs and the plant can increase in size, as long as sufficient water is present to permit the ions to move. The rate of growth probably is correlated with the rate of movement from one charged nucleus to another.

One of our better seed companies asked me to investigate a lima bean problem. A grower had bought enough seed to plant forty acres, but when the seed would not germinate, he sued the seed company for having sold him poor seed. His soil was a loamy sand with a low base exchange capacity. The seed company had sold some of the same seed to another grower who had harvested a good crop. I took several bushels of soil from the plaintiff's field and some of the same seed that he had used to plant his field to my greenhouse and I tested the soil. I found the pH reading near the neutral point; but the available calcium reading was less than 50 pounds per acre. The soil itself had an available calcium requirement of 2,800 pounds per acre-foot. I filled several eight-inch pots with the soil. In several others I mixed limestone with the soil, and in several others I placed a mixture of gypsum in the pots with soil. I then planted a dozen seeds in each pot, wetted them down, and waited for germination.

In four days germination had started in the gypsum and limestone treated soils. At the end of two weeks, eight plants per pot in the gypsum treated soil and eleven per pot in the limestone treated soil had formed their true leaves. With no treatment, four of the seeds had started to germinate but had not gotten beyond the cotyledon stage. In other words, the soil receiving the calcium supported quick seed germination; whereas the soil saturated with natural sodium instead of calcium would not support germination.

I have always been much impressed by the work done by Gedroiz in Germany many years ago. He conducted an experiment to determine the importance of saturating the soil with calcium instead of other positive ions. He took a soil and removed all the available calcium from it. One half of the soil he divided into different lots, and treated each lot differently with various salts to replace the calcium. He planted seed and got no germination except where he used a calcium salt.

The other half was mixed with 2 tons of limestone per acre, and the different lots were treated as in the first lot. Again only where he used the calcium salt did he get good germination. The limestone did not improve conditions for germination immediately. Then he set the pots aside for six months and again planted seed. This time seed germinated in all the lots which had received limestone but in the other series only the one with the calcium salt permitted seed to germinate.

The importance of calcium in arid, alkaline soils was brought to my attention many years ago when I was consulted on a citrus orchard in Arizona. I had previously published a paper on the relation of the form of nitrogen utilized by plants and the pH of the soil in which the plants were growing. In other words, ammonia nitrogen was most effective in soils of low acidity; whereas the nitrate ion gave best results on soils of high acidity. This was in soils having available calcium.

But when sulphate of ammonia was applied to Arizona's citrus trees, growing on the alkali soil, there was no response, and my theory was criticized. The foliage did not turn green. Nor did nitrate of soda produce a response. First of all, his problem was calcium, not nitrogen, deficiency. At the high pH the calcium in calcium carbonate was not available to the trees. Therefore, calcium nitrate and not nitrate of soda would be an ideal source of nitrogen because with the increase from less than 100 parts per million (p.p.m.) to over 3,000, the trees all turned green and what minor element deficiencies were present had disappeared. A combination of calcium nitrate and ammonium nitrate might have given an even quicker response.

LIME AND SOIL ACIDITY

Calcium in the soil is like grease on an axle. It smoothes out irregularities. One grower told me that the ease of plowing paid for the lime. During my college days, I had occasion to take many agricultural courses (at the expense of forgoing a liberal education). I was disappointed in many of these courses. First of all, no attempt was made to inform the students that the courses were only the application of scientific facts to crop production. Secondly, many of the people giving these courses gave out information in a parrot-like procedure without regard to proven facts. I noticed that these courses were on no higher level than those taught to us in the county agricultural school.

During the course work many platitudes were thrown out which meant nothing, covered up ignorance, and had not been and could not be proved. Some of these applied to the use of liming materials, things which I found I could discard without interfering with my accumulation of observations and facts on crop production. One of these was, "Lime makes the father rich and the son poor." I would change this to read, "Lime makes the father rich and the son a capitalist." The statement as it stands indicates that we know nothing about the action of limestone in the soil. It always seems to me that when we see the tremendous, unlimited tonnage of various forms of limestone piled up in our back yards as a result of natural forces, and consider the importance of the calcium ion not only in our crop production practices but in the health and well-being of our animals and human beings, we, as scientifically trained people, are neglecting one of the greatest God-given gifts to humanity.

The phrase "overliming injury" pops up every time any mention is made of the use of liming materials. It is a phrase that lingers on the lips of most people whose responsibility it is to hand out agricultural information. In itself it means nothing and, therefore, is a phrase of the uninformed. With multiple adjectives it could mean something very definite. It is a phrase that was added to our literature when wood ashes and burned lime were used on the soil to correct soil acidity.

Wood ashes have been used by plant-growing people since the eighteenth century. The story is told about the Mennonite scout who was sent out to look over a site for a settlement. When he inquired about an area adjacent to a river in one of the southern states, he was told the land was worn out. But he took a second look and found small round areas covering five to ten square yards where weeds grew in abundance. On examination he found that these plants had deep roots, while those nearby were very shallow. As he studied these areas, he surmised that they were the campfire sites of Indian tribes who had frequented the area many years before. Wood ashes and burned oyster shells had been left. He encouraged his people to buy this "poor" land and apply liming materials to grow their crops. Today this is a highly productive community of farmers.

I recall an interesting story published many years ago in *Reader's Digest* about a Reverend Mr. Orton, who was sent to the Smoky Mountains to take over a poor, run-down Methodist church. His first Sunday he had fewer than ten people to listen to his sermon. He decided something had to be done. These people were too poor to come to church. After a survey of his area, he decided he must show them how to raise enough food to at least fill their stomachs and in this way bring them to better health. Unlike his predecessors, who came to the church and left after the first sermon, he saw the light that would lead the community to better things.

He bought a piece of land adjacent to the church. He applied adequate liming materials to grow clover. He grew corn and other crops in abundance. The idea caught on and spread. Over a period of years following his experiment, the community prospered. He built more churches to accommodate the people. The health of the community improved. The use of limestone had brought health and happiness to a large community; and he called it religion. Perhaps we need more practical religion like that Dr. Orton handed out. We certainly have an abundance of limestone to do the job.

There is a reluctance in the general farm public to the use of sufficient liming materials on the soil. One would assume that a subject whose importance was brought to our attention several decades ago would be accepted without question in our day. That is not the case. Our farmers need many times more liming materials than they are using to get maximum yields. Not only are we short on lime in naturally acid soils, but natural limestone soils and those neutral soils in the southern states will respond well to liberal applications of liming materials.

A number of years ago the crops people in an eastern state and every department in the experiment station pooled their information on crop production and their efforts to find out what the main problems of the tomato crop were. After a survey conducted for three years in succession and involving over one hundred farms, the conclusion was reached that the more dollars a man spent for liming material, the more tomatoes he could expect to harvest from an acre. When the same study was applied to sweet potatoes under the same conditions, the same conclusions were reached. In other words, the insects, diseases, kinds and amounts of fertilizer, and soil types had only minor effects.

When one of the county agricultural agents told me that he wanted to initiate a program that would have some lasting value to his county, I mentioned the results of the survey and, since his county grew a lot of tomatoes with low acre yields, I asked, "Why not try to raise tomato yields?" His county was in a dairy state, and corn and alfalfa were a necessary crop. His farmers followed a four-year rotation. He catalogued his farmers and urged them all to apply at least 2 tons of pulverized limestone to each acre of tomato ground. In ten years he raised the average yield in the county from 3 to 10 tons. "But that is only part of the story," he told me later. "Our corn and alfalfa yields also increased as a result of the liming program."

I told one grower he needed 6 tons of limestone per acre. When he started to apply the limestone, he had some difficulty moving over the ground, and found he applied 6 tons on the first half acre. He said clover and alfalfa seeding was perfect on this strip, whereas he could see no response on the 6 tons per acre strip. The third year after he applied the limestone, the yield on the 6 tons was as good as on the 12 tons per acre strip. In many cases, the accidents tell us more than planned experiments. As a result of such experiences, I do not depend on the pH test to determine the need for limestone. Furthermore, I have seen cases where 2 tons of limestone and even hydrated lime applied to an acre actually decreased the yield, whereas 4 tons gave a substantial increase in yield. I have corrected many fields that seemed to have too much lime by adding more limestone. There is much research needed to find the real reason.

Many farmers don't want to be shown. In a meeting that I addressed, a grower told me he had tried everything and he knew you could not grow 100 bushels of corn per acre. When I asked him if I could work with him and find out whether he could increase his yield, his answer was an emphatic No!

I have conducted many field plots on farms where the increases were below the differences needed for experimental significance. Such plots don't help to solve farm problems. They do make you question your thinking. As a result, I have found that almost all soils need much more lime than a soil acidity test indicates. I have found that the reason limestone doesn't always show a response is because of such factors as inadequate mixing with the soil, prevailing moisture conditions, fineness of grinding, and kind of limestone—all affect the speed with which the calcium saturation is accomplished. Unless a certain saturation point is reached, yields will not be increased.

LIME CONTROLS PHYSICAL CONDITION OF THE SOIL

For many years we have considered lime a corrector of soil acidity. The soil acidity tester was standardized for an acre-foot or 6⅔ inches of soil. As long as commercial fertilizer was being used sparingly and barnyard manure was being used, problems concerning serious soil fertility deficiencies did not exist. There were carriers present in mixed fertilizers which could either acidify or sweeten the soil, but the quantities were so small that no serious problems resulted. Fertilizers with nitrogen from ammonia tended to build up more acid. Fertilizers with nitrate of soda or calcium cyanamid reduced the acid in the soil. Using both, one neutralized the other. However, the roots of crops explored the subsoil and removed calcium. Legumes, clover, and alfalfa feed heavily on calcium and magnesium. Removal of this calcium gradually tended to build up acid conditions in the soil by leaving acidic residue ions, unless there was enough residual limestone present and time was available to replenish the supply of available calcium. Limestone soils tended to maintain fertility over a longer period than the non-limestone soils, which had no residual limestone to draw on. Size of particles and solubility all contributed to the supply of available calcium.

Thus, when we build up acid in the soil or remove calcium, particularly in the non-limestone soils, we must apply limestone in adequate amounts. One ton of limestone will add from 400 to 900 pounds of calcium or the equivalent carbonate. Thus, for every 400 to 800 pounds of calcium needed in the soil, we must add a ton of limestone. If we need a ton in the surface 7⅔ inches, we may need an additional ton in the second layer and succeeding layers, particularly on the naturally acid soils. In order to speed up the effect of the limestone, it must be finely divided and thoroughly mixed with the soil, and placed in the subsoil by means of deep plowing or the use of a subsoiler. The effectiveness of liming material depends on the fineness of grinding and the thoroughness with which it is mixed with the soil. The action comes about by contact with acid particles. Being only slightly soluble, the calcium can be absorbed by the acid particles on the soil colloids in the base exchange complex only slowly. If strong acids are present, the solubilization of the limestone is accelerated.

Chlorides found in commercial fertilizers increase the need for liming material. When the chloride ion is released from muriate of potash, it must find something to attach itself to. Calcium seems to be a convenient companion, so calcium chloride is formed and, being very soluble, starts to move. If there is good drainage from the soil, the calcium chloride can be found in the drainage water in appreciable quantities. I have measured 40 p.p.m. This means that liming materials must be added to replace that which leaches away. For every 100 pounds of muriate of potash applied to the soil, 20 to 30 pounds of calcium is removed. A ton of muriate could conceivably remove the calcium from a ton of limestone.

When sodium or potassium salts (both common in mixed fertilizers) are applied to the soil, they are quickly dissolved in the soil solution and, being very active, they soon increase the calcium in the soil solution by replacement on the base exchange complex. Potash has been given the credit in many experiments for increasing yields, when actually the increased yields have been due to the increase in available calcium. Where a large supply of combined calcium exists, the application of other basic ions can stimulate growth. This is the reason why asparagus growers got good response to salt applications the first few years they applied it. When the calcium was depleted, salt no longer gave a stimulation. Many of the effects of nitrate of sodium could be attributed to the increase in available calcium rather than to a direct effect of the nitrogen.

With ordinary mixed fertilizers, where a low-grade superphosphate was applied, approximately one half of the superphosphate was gypsum or calcium sulphate. This calcium could offset the calcium lost with the chloride ions in drainage water. If, however, one of the high-analysis fertilizer mixtures, which are entirely soluble, is applied, the calcium problem becomes more critical, since there is no calcium sulphate present and all calcium must come from the base exchange complex.

We can prevent the loss of calcium by using sulphate of potash. When the plant removes the potassium, the sulphate ion also adds to the soil acidity; but when it combines with calcium, the resultant calcium sulphate is not very soluble and stays in the soil. This has other good features which will be mentioned later.

When we consider the above in terms of natural limestone soil, the story is somewhat different. Up to a few years ago, scientists assumed that limestone soils did not require additional limestone. If we depend on the soil acidity tester, we probably would find the soil to test neutral because small quantities of limestone would be in solution and could affect the soil acidity. Limestone soils will test neutral, particularly during the fall, winter, and spring months, when larger quantities of ammonia than nitrate nitrogen are present. When this ammonia becomes oxidized to nitrate nitrogen, we may get some high acidity readings. However, since most of these soils are tested during the winter months, the need for calcium is easily overlooked. Many of our soils seemingly have sufficient calcium when actually the available calcium is too low for good growth.

A number of years ago I worked with greenhouse soils for Yoder Brothers at Barberton, Ohio. When I first tested the soils I found the pH reading to be as high as 8.4, in spite of the fact that calcium deficiency symptoms prevailed on the foliage of the plants. The soil also was very compact, and became very slippery on the surface when wet, so much so that green algae grew freely on the surface. This usually means poor drainage and poor aeration.

My friends, Dr. Richard Bradfield, formerly at the Ohio State University in Columbus, and Dr. Barnes at the Wooster Experiment Station, and I discussed this problem on numerous occasions. It was puzzling. The crops were not producing and yet the soil test seemed satisfactory. The high pH camouflaged a lack of available calcium.

After considerable discussion, the need for a calcium test entered my mind. Dr. Bradfield and Dr. Barnes were skeptical of its value. The problem required more than a test for calcium. It was necessary, first of all, to find out how much calcium was necessary for a given soil, what the saturation point in the soil should be, and how this should be fitted in with our high pH. After several tests and experiments, I decided to pay less attention to the pH test. The potassium and sodium readings on the soil were excessive. I went back to the original source of information—as presented by Gans, Way, Hissink, and Gedroiz—which started me on the calcium test.

I filled a fifty-gallon cylinder with the soil which had a pH of 8.4 and leached it with distilled water. Every day I would add 5 inches of water, collect the amount that came through, and test it for calcium and potassium. At first, the potassium was very high, much higher than the calcium, but as I continued to apply water, the calcium and potassium leached out in equal quantities. The amount of calcium was very low. I applied 35 inches of water in all before the potassium and calcium decreased to the point where the amount equaled that found in an ordinary soil. The pH of the soil dropped to 6.8.

Checking back for the past several years on the treatment of the soil, I found that each year a ton of muriate of potash had been applied to an acre of ground for each crop. In the first few years that this had been done there occurred a good stimulation in yield, which prompted the growers to continue the practice. But the practice was continued until it no longer did the plants any good. The cucumber plants began to exhibit mosaic-like symptoms which became more and more common, as did calcium deficiency symptoms.

My interpretation of this was that after the first applications of muriate of potash, calcium was released and stimulated growth. This soil was of acid origin and contained very much organic matter. The colloidal complex including the decomposed organic matter formed a high base saturation complex, which could absorb large quantities of calcium. The calcium that was absorbed was soon replaced by the added potassium, since the base saturation of calcium never was very high. As the practice of applying muriate of potash continued, there was less and less calcium to be released. Finally it got so low that the soil solution did not have sufficient calcium to antagonize the potassium ions. Then calcium deficiency symptoms began to show on the plants. The nutrient ions were thrown out of balance and trouble showed up for the grower. I have found this to be a common problem in many areas where the application of lime has been neglected.

I immediately brought in several carloads of dolomitic limestone and one carload of Youngstown slag and applied it to these soils at the rate of 10 tons per acre. It took a whole crop before the full effect was noticed. A rototiller was used to mix the material with the soil. Root growth was normal, the pH dropped to 5.4 and gradually came back to 6.9, deficiency symptoms, including mosaic-like symptoms, disappeared from the cucumber and tomato leaves, and the soil became mellow. Drainage was greatly improved. Of course, this was a shotgun method of doing a job, but it meant profits to the growers. The pH was incidental to the problem.

As a result of this and other similar experiences, I have depended less and less on the acidity test for determining calcium needs. I feel that agronomists are handicapping their work by placing too much confidence in testing for acidity soils where mixed fertilizers are being used. In many problems that I have worked on, growers have shown me soil reports in which the pH was 6.8—no lime recommended; and yet I applied 4 to 6 tons of limestone because the available calcium was too low. I have increased yields from 50 bushels to 165 bushels by applying sufficient limestone and no additional fertilizer.

One of the best co-operators I have came to me for advice in 1952. He had decided, after serving in the war years with the Navy, to take over 80 acres of land belonging to his father's family to try to make a living for himself and his bride. He secured all the advice he could. According to the pH test he did not need any lime. He was advised to use 700 pounds of mixed fertilizer an acre to grow corn. He started to farm.

Before the war he had built up a flock of sheep as a 4-H Club project. He decided to raise sheep and started another flock. At the end of four years he was unable to grow over 50 bushels of corn on an acre and his fertilizer bill was making it impossible to show any profit. When he inquired about this, his advisers told him that his land was submarginal, and that if he wanted to grow more than that he would have to buy better land. This seemed good advice.

By accident he and I got into a discussion of his problem. To me, it was a challenge, and I asked him if he would like to do some experimenting if I would furnish the fertilizer. He agreed to go along.

I tested all his fields on the basis of available calcium and found he needed 10 tons of finely ground limestone on every field. The land was rolling and variable in composition. Some was river-bottom land. Some contained much clay, silt, and gravel. He did not have enough money to buy limestone for the whole farm, so we decided to take a 16-acre field as a start. Limestone was applied and the field was planted to corn with 3, 6, and 9 gallons of 7–14–7 fertilizer solution. The field was plowed, disc-harrowed once, and planted. A rotary hoe was used on the corn crop once. Weeds were killed with weed killer. Up to this time, this field had never grown over 40 bushels of corn per acre.

On the 16-acre field the average yield the first year was over 100 bushels per acre. The remainder of the 80 acres in the farm was later given the same treatment, with equally good results. A pH test on such soils does not help very much because of the strong buffer system that exists. It is necessary to determine the degree of base saturation and calculate the amount of calcium needed to do the job. (I have talked with growers who told me they applied 3 tons of pulverized limestone per acre and found the pH reading lower than it was to start with. The first impulse is to say that the limestone is of no value. A lowering of the pH may be due to the time the test was made or the soil's being too dry or wet. The fact that the limestone did not raise the pH could very well have been due to a buffer system which might require more than 3 tons of limestone.) Five years after I started working with this grower, he won the local 100-bushel corn contest with 143 bushels of corn. The crop was grown with 2 gallons of 10–20–10 fertilizer solution in the row and 2 gallons of the same material sprayed on the foliage two weeks before the corn was ready to tassel.

The pH test is standardized on an acre-foot of soil. In other words, we determined the limestone needed to bring the soil up to the neutral point and not to 85 per cent saturation of the base exchange requirement. It is greatly affected by our fertilizer program—residues left in the soil. However, we know that for the best results the roots must penetrate deep into the subsoil to be assured of an ample supply of water. Therefore, our lime correction problem is to add enough limestone not only to supply the plowed layer but to supply numerous additional layers, which may mean augmenting the 8-ton surface application by 2 to 5 or more tons. Also, we must keep in mind that some soils are formed from acid rocks which are acid to considerable depths. Soils formed from limestone rocks may need limestone only in the surface layer. But even here the pH test may not be of much help. A calcium test should be made. I have tested many soils from limestone areas that were neutral, but the available calcium reading was so low that an application of 3 to 6 tons of limestone was necessary.

I tested a soil for one grower and found the surface soil adequately limed. The grower planted corn, and up to the silking stage, it looked like 100 bushels of corn to the acre. When he picked the crop, he had less than 20 bushels per acre. The ears never did grow. When I checked the field again, I noticed the roots were all in the surface 6 inches. The bottom of the furrow was so hard it was difficult to get a good sample. When we tried to germinate corn seed in it in the laboratory we had no success, but when I mixed a teaspoonful of limestone with a coffee can full of the soil and planted corn, the germination was above 90 per cent.

It seems to me that if we can double the yield of corn in a field by applying 3 to 4 tons of limestone when we can't get a response to additional fertilizer, the problem involves the chemical and physical condition of the soil, not the amount of fertilizer that we apply. We must keep the horse before the cart. We have been keeping the cart before the horse.

Observations on our experimental farm at Olena, Ohio, and observations made by farmers who follow our recommendations, convince me that any soil that can be farmed can be made productive by applying adequate amounts of pulverized high-calcium limestone. The same may be accomplished with dolomitic limestone, but the amount to apply must be figured on the basis of its calcium content, not on its total neutralizing value. In other words, if it requires 8 tons of high calcium 45 per cent limestone, it would require 14 tons of a dolomitic limestone having 25 per cent calcium.

The final effect of the application of this limestone, according to many observations, is multiple:

1. It requires less horsepower to pull a plow through it.
2. It mellows the soil to much greater depths.
3. It improves drainage and speeds oxidation.
4. It more than doubles yields.

Different types of soils require varying amounts of limestone.

BLACK SOILS NEED MORE LIMESTONE

A New Jersey celery grower, Mr. Anderson, discussed his soil fertility problem with me. He grew celery on some of the black bottom land along a tidewater stream in central New Jersey. He said, "I have had an experience with lime on celery that doesn't make sense and now I want to know whether I am headed in the right direction." Then he told me his story. "Four years ago my celery was hardly worth harvesting. I had the soil tested and was told I needed three tons of limestone." (A mere drop in the bucket, I thought.) "My celery wasn't improved much. I spent the winter in Florida running a fishing boat. On one of the trips, I happened to talk to a man who did research work on soils in the Department of Agriculture in Washington. When I told him about my soil and what I had done, he told me that I probably needed a lot more lime. Then he added, 'If you will send me a sample of soil, I will tell you what I think,' which he did." Before Mr. Anderson started back to New Jersey, he received a letter from the man in the Department of Agriculture saying that he would need at least 9 tons of pulverized limestone and perhaps even more. The grower said he took the letter to the man in the experiment station who had run his first test. When he read it, he said, "My God! You will get your pH so high that you will overlime the soil and hurt your celery." He said, "I couldn't hurt that celery any more."

The grower went on. "I thought this over, and since I could not hurt the celery any more, I took a chance and put on another three tons per acre. That year my celery was better but not as good as I had grown before. So the third year I put on the third three-ton application—and you never saw such a crop of celery. The fourth year I figured that I had enough limestone so I didn't put any more on and, you know, my celery wasn't as good. Since then, I have been applying a ton every year. You know, I had a pH test run on that soil and it is not above the neutral point."

When he asked me why that was, I told him that we had a lot of limestone soils in the United States that contained 50 to 150 tons of calcium carbonate (limestone) per acre where the pH was never above 7.0 because of the limestone. If he had used hydrated lime, which is much more soluble, he probably would have run the pH to 8.6—and had worse trouble with his celery. I told him that I recommended limestone freely, but that I recommended hydrated lime with a great deal of caution. I told him he could get good results with hydrated lime if he used it often and in small quantities. I have applied as much as 40 tons of limestone per acre on some fine, black soils high in organic matter before achieving maximum yields.

During my lifetime I have been called a lot of names because of my adherence to my ideas on soil fertility. People have called me "lime crazy," "the man who has limestone running out of his ears," "the lime dictator," "the lime man who is paid by the lime companies." As long as limestone pays as good dividends as it has for me, I shall continue to recommend it. I do it because the farmer can make money using fertilizer only when he has enough limestone in his soil. I see no reason for spending money for fertilizer if a farmer can't make money using it.

I am convinced that if a farmer uses adequate amounts of lime on his land, he will be rewarded by far greater profits than he can expect from any other practice. I have told many farmers who did not have sufficient funds that they would be far smarter to spend all the money they had for limestone, not for fertilizer, until they were sure limestone was no longer necessary. Then they could expect some big profits using fertilizer.

ACID-LOVING PLANTS AND LIMESTONE ARE COMPATIBLE

Many of our textbooks contain tabulated lists of plants showing lime needs based on the pH requirements or acidity of the soil. In most cases, these lists represent groups of plants which should be grown at pH 4.5 to 5.0, 5.0 to 5.5, 5.5 to 6.0, and 6.0 to 7.0. There is only one real interpretation for such data. It shows that some plants will grow at a pH as low as 4.5 and others can be grown only above 5.5 and others must have a pH not lower than 6.5. In other words, it means tolerance to acid conditions or low calcium saturation of the exchange complex. This type of data has given rise to the idea that there are acid-loving plants, that if on a neutral soil you wanted to grown beans which according to these lists should be grown on a soil having a pH of 5.5, you would have to add sulphur to increase acidity. Then, if, after beans, you wanted to grow spinach, which requires a neutral pH, you would have to add large quantities of lime. How ridiculous this thinking really is! We know that even though beans will tolerate an acid soil, they will do much better when grown on a neutral soil. The important thing is to get the soil in the proper chemical and physical condition. A good chemical condition means a good physical condition.

The Azalea Gardens of Norfolk, Virginia, were having considerable calcium deficiency symptoms on azaleas. They were losing plants every year. The superintendent asked me to help him. We (Dr. L. L. Danielson and I) took some of their soil and sick plants to our greenhouse and set up an experiment using two-gallon glazed pots. Since the soil was very acid, we applied some dolomitic limestone to several of the crocks—at the rate of 2, 4, and 8 tons per acre, mixing it thoroughly with the soil—and set the sick plants in all the pots. After several months, we noticed only those with 8 tons of limestone were growing rapidly, although past experience indicated that these plants should be grown on strongly acid soil.

Dr. Danielson, our plant physiologist, was interested in the problem. I asked him to lay out a series of field plots 50 feet by 20 feet, and we agreed on the following treatments:

1. 100 lb. sulphur per acre	This brought the pH of the soil down to 4.7—strongly acid. The textbook recommendation.
2. 50 lb. sulphur per acre (recommended as good practice)	This soil had a pH of 4.9.
3. Nothing added	pH 5.4.
4. 2 tons pulverized dolomitic limestone per acre	pH 6.5.
5. 4 tons per acre	pH 7.2.
6. 8 tons per acre	pH 7.2.
7. 16 tons per acre	pH 7.2.
8. 32 tons per acre	pH 7.2.

Regardless of the amount of limestone applied, the pH was the same on all the limestone plots beyond the 2-ton application.

We planted rooted cuttings, and small plants of many different varieties that supposedly required an acid soil, in rows across all the plots. We had six varieties of azaleas and one variety each of camellias, gardenias, roses, tung oil trees, blueberries and others. On Plot 1 nothing grew after the first year, partly because the plants made such shallow roots that heat and dry weather killed them. We made no attempt to mulch them. We applied no mulch to any of these plants, even though it was a common practice to do so. Some of them made a little growth, but they all died after the first winter. There is a lesson on mulching here. The general practice is to mulch these plants because they are considered shallow-rooted plants. They need mulch because they practically grow in the mulch. However, on Plots 4 to 8 they rooted deep into the subsoil and needed no mulch. In three years, one rooted gardenia cutting 3 to 4 inches long grew to be a plant 4 to 6 feet across with beautiful dark-green foliage. Azaleas and rhododendrons grew as well on Plot 4 as they did on Plots 7 and 8. In general, they all grew much better with 2 tons of limestone, but they were not hurt by 16 to 32 tons of magnesium limestone per acre. There was no chlorosis on the foliage of the plants grown on the limestone plots. There was considerable chlorosis on the plants growing on the first three plots.

Had we used hydrated or burned lime or wood ashes instead of limestone, we probably would have had difficulty keeping the plants alive through the first winter because of chlorosis. When I published these observations in *Horticulture* magazine, I received a letter from a scientist criticizing my statements and observations. Among other things, he mentioned that "ferns do not grow on the limestone bluffs in Eastern Missouri but rather in the valleys below the bluffs." I answered him by saying that the reason for this was the fact that all the available calcium was in the valley and there was none on the bluffs. I think many people misinterpret what they see, that many ideas we have were handed down to us by people who did not understand or misinterpreted the facts. The idea that you can't use limestone on certain plants comes from the use of wood ashes. The burned lime in wood ashes is too quick-acting for many plants. It creates an abundance of free calcium which prevents the plants from absorbing other needed nutrients, and they show symptoms which may be called overliming injury. However, that is no proof that they won't grow with limestone. Pulverized limestone is much less soluble than burned lime and needs a growing season before a proper balance of nutrients occurs.

Several years ago I told my wife to order some pulverized limestone and to apply it freely in preparing a bed where she wanted to set out chrysanthemums. I warned her not to use hydrated lime, which is the burned form of limestone. I told her to use the limestone liberally, which she did. When I got home that evening I saw a bag of hydrated lime sitting near the bed. The garden center had sent her the wrong material. I immediately turned water on the bed, hoping to keep the plants from dying. They were all wilted. We saved about half of them, but it took three weeks before they showed signs of recovering. That season they only produced a single stem with a single flower. Even the second year they did not grow as they should. The next year they grew beautifully, as did other plants that were set in that year. It took two seasons for the harmful effects to wear off.

The danger of hydrated lime is its activity. It should be applied the fall before you wish to start a garden. If this practice is followed, the calcium has a chance to become carbonated and also absorbed by the soil colloids, after which it is in equilibrium with the soil. It will no longer interfere with the proper functioning of the roots nor prevent the absorption of a balanced diet of mineral nutrients.

I have had occasion to try out the use of magnesium pulverized limestone on azaleas and gardenias in many landscape plantings, and successfully prevented chlorosis where most people were trying to correct with iron sprays.

In some cases these plantings appeared to be growing in poorly drained locations. The plants were lifted, limestone was mixed liberally with the soil, and the plants were reset with excellent results.

I remember one case where a clump of azaleas was growing on high ground, but the foliage was always chlorotic, as though the drainage were inadequate. We lifted the plants and found they had been set in with sphagnum moss. This was very wet and the roots had made very little growth. We mixed a gallon pail full of limestone with the moss and dug up some soil to give a 1-part-moss to 4-parts-soil mixture, and set the plants back in the same place. It took almost a year before the plants turned dark green and started to grow. They developed into beautiful specimens.

There are many other plants included in the "acid-loving" group with which I have worked which have responded to limestone treatments. Strawberries have been considered to belong to this group, and yet the best strawberries I have seen grown were on soil where enough pulverized limestone had been added to satisfy the calcium requirement.

Blueberries, definitely acid soil plants according to authorities, will do much better, according to my experience, on soils where the soil is well treated with magnesium limestone. Mulch is important for blueberries. Gardenias supposedly do best on very acid soils, but I have never experienced this. The most rapid growth, very green color, that I have seen, was grown on soils to which eight tons of magnesium limestone had been applied. The amount of limestone needed depends on the type of soil. A sandy soil needs much less than a heavy soil because it has less colloidal clay and chemically active organic matter.

I have tried to determine where our ideas on the need for acid soils originated. Acid soils limit the amount of vegetation per square yard. Plants that are tolerant to acid conditions meet less competition on acid soils. I have seen rhododendron growing in mountainous, wooded areas, with no competition because of the very acid soil. And yet I have grown beautiful rhododendron on soils that were heavily limed with magnesium limestone. Apparently, because of the tolerance of these plants, we have assumed that they must be grown on acid soil; whereas they will grow much better on limed soil, if they don't have to compete with other species.

I have seen wild strawberries completely cover acid gravelly knolls. There was no competition. I have transplanted these plants to my garden where the soil was sweet and fertile. They grew well, much larger than in the place where I found them, but they did not produce berries and died out in a year or two because wild clover and perennial weeds crowded them. If they were carefully weeded, they made beautiful plants and grew from year to year, but the berries were sour and fewer per plant. Apparently, too fertile a soil was not suited to their continued existence. They had become adapted to soils of low fertility and acid conditions.

CHAPTER 3

Some Crops Are More Sensitive to Calcium Needs

CALCIUM IN RELATION TO TOXICITY OF SPRAYS AND FUMIGANTS IN CONTACT WITH THE FOLIAGE

THE importance of calcium in building up protoplasm resistance to the toxicity associated with certain sprays and fumigants, and its relation to the killing effects of herbicides, are too often overlooked. The following story emphasizes the importance of people with different training working together.

The importance of pulverized limestone in the soil to the general welfare of cucumbers, as previously mentioned, was of much concern to the owners of a cucumber-producing greenhouse plant in Barberton, Ohio, who prompted me to initiate several pot experiments. The soil was known to have a high pH of 8.4, with a very low reading of available calcium. (I want to give much credit to Dr. Barnes and Dr. Bradfield, who were with Ohio State University at the time—1932–1934—for their stimulating ideas and discussions helped me greatly in formulating these experiments.) The high pH of the soil, along with a very low available calcium reading, were difficult to understand in terms of our ideas on the reliability of the soil acidity test in determining lime needs of the soil. (Since the publication of the 1957 U.S.D.A. *Yearbook*, we have a better understanding of this.)

The potassium content of the greenhouse soils was very high, due to excessive applications of muriate of potash, a ton to the acre having been applied every year. This undoubtedly had much to do with upsetting the soil nutrient level. Much of the calcium leached away as calcium chloride.

To set up the experiment ten quart tubs were filled with soil and were separately treated with different amounts of pulverized limestone. Successive tubs except the check received the equivalent of 400, 800, 1,200, 1,600, 2,000, 2,400, 2,800, 3,200 and 3,600 pounds of calcium per acre. Each tub treatment was repeated four times. Individual cucumber plants were grown in each tub and supported on strings hanging from a wire 8 feet above the tubs. There were differences in rate of growth, from the check plants, which grew slowly, to those receiving 2,800 pounds of calcium, which grew more rapidly. Beyond that, there was little difference in the rate of growth.

When the first plants reached the overhead wire, some of the margins of the older leaves on all plants which received less than 1,600 pounds of calcium per acre began to turn yellow and die. This marginal burning is often mistaken for potassium deficiency. When the plants had cucumbers ready to pick, sulphur dioxide, from sulphur which had accumulated on a six-inch main line steam pipe which was used once a year to carry steam for soil sterilization to a greenhouse beyond, was accidentally released in the compartment.

The next day many of the plants were entirely dead; whereas those receiving 2,800 pounds or more of calcium showed no noticeable injury. When the damage was evaluated, all the plants receiving 1,600 pounds of calcium per acre or less were dead. Those in the tubs having between 1,600 and 3,200 pounds exhibited considerable damage to the older leaves. The results are shown in the accompanying figures. Apparently, the injury was indirectly correlated with the amount of available calcium in the soil. Several years later, I was discussing this with Mr. Fuller, who marketed the Fuller method of greenhouse fumigation to kill mites on flowering plants. He said he could not understand why his method seemed perfectly safe in some houses while in others it did considerable damage. As he thought about it, he said he had no difficulty in well-managed houses. Injury occurred in badly managed greenhouses. I related my experience with cucumbers and told him to check on the available calcium in the soil. Perhaps he could find the answer. Several years later he told me he had restricted his fumigation to greenhouses that were well managed and where applications of pulverized limestone had been made.

In 1934, after I returned to New Jersey and started a research program on soil fertility problems, I reported results on a pH–available calcium problem in *Soil Science.* We were finding many similar cases in sandy loam soils due to excessive uses of nitrate of soda in the production of vegetable crops. During the next twenty years I ran into this same problem in many different areas east of the Mississippi River.

Some six years after I had had the experience with the cucumbers, I was asked to work on a co-operative project where arsenic injury was being studied on fruit trees in New Jersey. The leaves on these apple trees had turned yellow and dropped off at about the time the fruit was half grown wherever arsenate of lead had been used for the control of worms in the fruit. Eventually the trees showed many naked branches with only two or three leaves on the tip. This condition was not unlike the symptoms of magnesium deficiency on apple trees. In the following year or two, the trees did not set fruit, and some of them died. Since arsenate of lead was a common spray ingredient, and since the foliage turned yellow at about the time arsenate of lead was used in the spray, this ingredient was viewed with suspicion, and chemical studies were started to find out how the arsenate of lead was causing the injury. It seemed that the injury occurred a week or two after the spray was applied. It followed the pattern of a systemic disease: no burning or immediate injury, but a gradual fading of the green color and abscission of the leaf.

After six years of study, trying to find out how arsenate of lead was doing the damage, we felt we were up against a stone wall. Nothing definite had been learned. At this time it was decided that the assistant extension specialist in horticulture, Mr.

Harold Robertson, and I were to make a survey of the orchards and find out how widespread this damage really was and whether we could find some correlations in the field.

From my previous experience, I was prompted to take a portable acidity tester with me. After visiting at random ten orchards, all of which were being sprayed with arsenate of lead, we found orchards ranging in injury from those in which trees were in poor growth with some trees dying, to orchards which were in perfect condition and yielding heavy crops of fruit. It was also noticeable that where the trees were badly damaged, cover crops would not grow very well under the trees. It was evident that arsenate of lead was not the real cause, although it did not eliminate the possibility of some indirect effect, since we found no orchards where arsenate of lead had not been used. We decided to investigate one of the most badly damaged orchards, which happened to belong to Paul Burke, on Rancocos Creek in Camden County, New Jersey. We found him very co-operative and anxious to work with us.

I must digress for a moment to give some background information, because sociological factors are sometimes tied in with cultural practices. To my way of thinking, Paul Burke was a gentleman fruit grower. He worked very closely with experiment station people, read, in addition to other things, everything he could find on fruit culture, and tried to do the right thing. He and his wife lived in one of the beautiful old homes in New Jersey, surrounded by antique furnishings which would do credit to many museums. Their family consisted of three sons, two in college and the third getting ready to attend Cornell University. The eldest son had attended the University of Pennsylvania and was the current Olympic sculling champion. Everyone worked, and it was very discouraging to see acres and acres of orchards gradually dying, apparently in spite of following recommended practices. As we walked through the orchards and saw the poor crops, our conversation revolved around the idea that a good crop of apples on such a fruit farm should pay the expenses necessary to assure three boys a college education; whereas a poor crop could actually just be an additional expense.

I resolved that I was going to solve Mr. Burke's problem if it was the last thing I ever did. I asked Paul what the lime condition in his soil was, and he said the pH was satisfactory, 6.4 to 6.8. The soil was a loamy sand and had produced exceptionally fine fruit in past years. As we walked through the orchard, we found spots near trees where some sweet clover plants were growing two feet tall. I grabbed a plant and was surprised that it could be pulled up with very little effort. When I examined the roots, I found that the tap root had grown one inch and had then divided so that it resembled an inverted Y, with the branch roots all growing parallel to the surface of the soil at the one-inch depth. Mr. Burke told us he had applied 2 tons of limestone per acre before the sweet clover was sown. He had disced the limestone into the ground. I got my acidity tester and checked the pH and found the soil tested 4.7. Paul said my soil tester was wrong, that he had tested the soil with his tester and it was 6.4. Then I took a sample of the surface inch of soil and we both got a 6.4 reading. The limestone was all in the surface. When he told me that he had run the disc harrow eight inches deep to mix the limestone with the soil, I told him he used the wrong tool to do the job.

We secured a shovel and started digging holes around the trees. All the older roots had sent feeder roots up to the surface inch of soil. Every time he disced the soil, he cut off all the feeding roots. We realized that the trees were starving. He used sulphate of ammonia as his nitrogen fertilizer. This was making the soil more acid.

When we realized that the problem seemed to be associated with a lime deficiency in the subsoil, we suggested that he apply limestone, 2 tons to the acre, plow it under, and put 2 tons on after the ground was plowed. Up close to the trees where he could not plow we suggested that he spread six to eight shovels of limestone. He was worried about cutting off the roots when he plowed the ground. I told him it would not be any worse than cutting them off with the disc harrow. Several months after he applied the limestone we dug holes around the trees again and found the soil full of feeding roots. That late summer I left

Rutgers University and I did not see this orchard for three years. When I did see the orchard again, it was producing a fine crop of fruit. I could hardly believe that this was the same orchard, and Mr. Burke informed me that he was still using arsenate of lead.

A number of years later I had occasion to work with a peach grower in one of the southern states along the Atlantic seacoast. This grower had 60 acres of fruit. When I first visited this orchard, the grower was alarmed about the growth condition of his trees, particularly since he had been told that he was not in a peach-growing area and that weather conditions and spray materials were responsible for the sickly appearance of some of his trees just when they were beginning to produce fruit. As we walked through the orchard, he pointed out trees that were showing signs of injury. When I asked him whether he had used limestone on the soils, he informed me that he had been warned by his college advisers that he should not use it as it might ruin the orchard. When I told him that he would lose a number of his trees if he did not put on limestone, he started an argument. I told him I wasn't interested in arguing, but that if he was willing to put on 3 to 4 tons of limestone per acre around some of his sick trees, I was sure the trees would be revived.

When he saw how much good the limestone did for these trees, he put limestone on all the trees. After that he had vigorously growing trees that yielded quantities of high-quality fruit. The college advisor still warns him against using limestone on peaches. The grower, for some reason, did not tell him about his putting on the limestone. The human equation is hard to understand at times.

It seems as though every time I mention limestone to a grower he tells me that he has been warned by his county agriculture agent not to use limestone. A number of years ago I told a spinach grower that his problems were due to insufficient lime in his soil, and he told me that his county agent had told him to be careful not to overlime. I set up some plots applying 2, 4, 6, and 8 tons of limestone per acre. I found out later that several people from the experiment station had taken pictures of the plots because they were sure that I would "overlime" the soil. The grower told me that when the spinach on the 8-ton plots began to grow better than on the other plots, they stopped taking pictures. The plot outyielded the others. I couldn't understand why they weren't interested in growing better spinach, and why they didn't take pictures up to harvest. Their attitude seemed to be to try to prevent growers from growing better crops rather than to help the farmer to do a better job. It was a case where the book could not be wrong.

We have too much of a negative approach to our fertility problems. A lot of research people—I should label them testers—seem to try to disprove anything that is new. They make up their minds that the new idea is wrong and won't work, and then they try to prove it. And if they can't prove it is wrong, they blame the weather. They would do the farmer much more good if they would approach a problem humbly and open-mindedly, and reserve their final opinions until all the evidence was in.

I have heard agricultural research people criticize people engaged in fundamental research in other fields as being long-haired and too impractical or so technical that nobody could understand them. I immediately classify such a person as ignorant or too lazy to try to understand. It is my candid opinion that our agricultural problem, if there is one, can only be solved by men who are steeped in fundamental research. There is no place for the politician in this picture. A farmer, to maintain a satisfactory standard of living, must look to the fundamental research man for guidance. Superficial thinking is responsible for low average yields, which can only lead to a low standard of living. The picture is not a pretty one; and our extension service set-up must assume a lot of the responsibility for making the picture as dark as it is.

Agriculture is complicated business. It is 100 per cent a chemical phenomenon and it takes chemical knowledge to understand it. We find farmers who are doing an exceptionally good job who have no chemical training, but for every one like that there are ten or more who are barely existing. To them, chemistry is bunk. I knew a college professor who, when confronted by some statement he could not understand, turned it aside by saying, "It's the bunk." He even wrote a book which was a repetition of what others had written before him in other books. People with such points of view should not be in a position where they can teach others. They are responsible for much of the agricultural misinformation that is disseminated for the farmer's use. It will be corrected eventually, but in the meantime many farmers will lose their farms.

THE AMOUNT OF CALCIUM IN THE SOIL AND THE GROWTH OF CUCUMBERS, TOMATOES AND CELERY IN A GREENHOUSE, AND CELERY AND HORSE-RADISH IN THE FIELD

During the early 30's I was employed by a large greenhouse grower in the Akron area to help him find out why cucumbers and tomatoes were growing so poorly with what seemed like ample fertilizer. The cucumbers grew to the wires 6 feet above the ground with much yellowing of the older foliage, which soon caused premature drying of the old leaves and much malformation of the fruit. Diseases seemed to be prevalent in abundance. There were many "nubbins," mature cucumbers not over 3 inches long. By this time the growing tips showed symptoms of mosaic.

The tomato plants seemed to grow freely enough, but they did not set fruit readily and much of the fruit that did set developed into rough, misshapen specimens. The leaves showed many chlorotic areas and premature drying of the older leaves.

An examination of the foliage with tests applied while examining leaves under the microscope showed a large amount of potassium but no available calcium crystals. When we examined the soil there was no available calcium. However, the pH of the soil was above 8.0. The potassium was very high and the phosphorus was low. A situation existed here which was contrary to general knowledge—a pH above the neutral point but a negative test for calcium.

The soil indicated a highly dispersed physical condition—very slippery and slimy when it was wet, and baking as hard as a brick when it was dry. The soil between the rows, where there was much traffic, was as hard and smooth as an asphalt highway. When it was worked between crops, it was hard and lumpy. It was very difficult to steam-sterilize the soil because of this lumpy condition.

It happened that a Dr. Doolittle, from the U.S.D.A. Department of Plant Pathology, stopped by about this time, so I had a chance to discuss our mosaic problem with him. When he looked at the plants, he asked, "Why are you applying so much potash?" I told him that I was unaware of any heavy applications of potash having been made. However, this agreed with the microscopic examination I had made previously. When we inquired about this from the grower, he said he had applied a ton of muriate of potash per acre before each crop of cucumbers, because he had been advised that if you wanted to grow high-quality cucumbers, you needed an abundance of potassium. He further informed us that the first time they used it the cucumbers were definitely better than the previous crop, but that after that succeeding crops were not of high quality. I assumed from this that the potash had made available, through displacement, liberal quantities of calcium the first few times it was used, and that succeeding applications released less and less calcium, which was not sufficient to balance the liberal quantities of potassium in the soil. This also could account for the high pH, because of the greater activity of the potassium ion. In other words, with no available calcium in the soil, the plants absorbed potassium in large quantities. There apparently was so little calcium and so much potash that the plants looked as though they had a disease. The soil (normally a good silt loam) was hard. Limestone could soften this soil; but the pH was above 8.0. (I later found that by adding magnesium limestone, the pH of 8.0 dropped to 6.8.)

I immediately got some of this soil into the laboratory, mixed it thoroughly, divided it into four lots, and filled eight-inch pots with it. One lot I put in pots with no additional treatment, for a check. To the second lot I liberally added pulverized limestone. To the third lot I added calcium sulphate, or gypsum, and to the fourth lot I added potassium nitrate. I planted cucumber seed and grew the plants on strings until the largest were 6 feet tall. Without any treatment, the plants grew slowly, and resembled the plants in the greenhouse. The gypsum and limestone plants were beautiful by comparison. They looked like well-grown cucumber plants. If there was any difference, it was in favor of the limestone. But the potassium nitrate plants were a sorry sight. They grew slowly and resembled the greenhouse plants, except that they were not as good.

When I took the plants out of the pots, I found that the roots on those receiving limestone were all through the soil, so that the soil fell apart when I removed it from the pots. The soil in the gypsum-treated lot was not as loose. The soil in the other two lots was hard. The roots had grown between the pot and soil and the soil held together firmly in a hard ball. When I tested the soil, I found it to be 6.8 in the limestone series, 7.3 in the gypsum series, and over 8.0 in both the check and the potassium nitrate–treated soil. I must point out here that I did not apply equivalent amounts of calcium as limestone or gypsum.

I decided you could have a high pH soil and still have calcium deficiency. I immediately ordered two carloads of dolomitic, pulverized limestone and spread 80 tons on the eleven acres of greenhouse beds. We used a rototiller to mix it with the soil. It was the end of our troubles. The so-called mosaic on the cucumbers disappeared, the spotty condition on the tomato leaves disappeared, and the soil became mellow instead of turning up in large, hard lumps.

I was asked to come back to the New Jersey Experiment Station after this, and the first problem I got involved in was a high pH, calcium deficient celery soil. When I told my colleagues what I had concluded, they said, "You can't have such a condition out here. You can only have that in the alkali soils." But when a grower brought me half-grown celery plants with the heart leaves rotting, I immediately said it was calcium deficiency. He said that it couldn't be because the pH was neutral. When I checked the plants and soil for calcium, I found none. There certainly was no calcium available to the plant. It was too late to do anything in the field, so we got enough of his soil to set up an experiment in the greenhouse. We compared the untreated soil with the same soil to which we had added limestone, and set in some of his sick plants. Without limestone, the plants made no further growth. With limestone, the plants started to grow and finally outgrew the calcium deficiency injury. When I tested the soil in the two lots, that with the limestone had a pH of 6.4 while the check lot tested 6.9. The pH had actually been lowered by the lime treatment.

I have seen this happen on numerous occasions. I have recommended 3 to 8 tons of limestone per acre on soils that had a neutral pH but very little available calcium, and have had the growers call me and ask me why their pH dropped to 6.2. They were always ready to condemn the limestone, but when we checked the soil for calcium we found it adequate for the soil type. I must warn anyone who conducts these tests that the pH will vary from 6.0 to 7.0 during a twelve-month period.

Soluble salts tend to move up and down in the soil, depending on its moisture content. During the summer, except after very heavy rains, soluble salts may be very high in the surface inch of soil. During the winter these salts are very low, accompanied by some leaching. During the summer, loss of nutrients occurs mostly from surface runoff. The soluble salts in the surface usually have very little calcium, unless the soil is saturated with calcium. Most of the calcium probably is lost by leaching. Water running out of drain tiles where large amounts of mixed fertilizer had been applied has been known to carry 40 p.p.m. of calcium, in the form of calcium chloride.

When I asked the celery grower how his soil had reached this low calcium condition, he told me that the farm had originally been a potato farm where the pH was maintained at 5.5 or less to control scab. However, the owner had found too much scab and had sold the farm. The present owner had grown a fairly good crop of celery with 150 pounds of nitrate of soda the first year. During the following years, he had found that he had to use more and more, until for the present season he applied 1,500 pounds of nitrate of soda, and his celery developed calcium deficiency. My explanation to him was that the soda probably was kicking calcium out of the exchange complex until most of it was replaced by sodium. The nitrate nitrogen probably did not help the celery much. The problem, therefore, was to replace this sodium with calcium. Since the calcium requirement of this Collington sandy loam should not be over 2 tons of pulverized limestone per acre-foot, it should have been a simple matter to correct. However, when we went to the field with the problem, we found a plow sole 2 to 4 inches thick under the plowed layer. In some cases this was as hard as concrete. So, we plowed under a ton of limestone and applied another ton on the surface and mixed it as well as we could. We worked on this problem for seven years, during which time we had applied 6 tons of pulverized limestone per acre; and our celery still suffered from what seemed like calcium deficiency. I finally asked Dr. Joffee from the Soils Department at Rutgers to work with me on this problem. He very carefully examined the soil profile to a depth of four feet, tested various layers and, after some calculations, told the grower he probably would need another 6 tons of limestone. He found later that the irrigation water, which came from a 300-foot, ten-inch well, contained an appreciable amount of sodium chloride. A new well was drilled 100 feet deep to give salt-free water. Nevertheless, applications of limestone gave a definite boost to the celery for several years after this. During this trial period, a smaller field where he grew plants developed calcium deficiency. When the plants were four inches tall, the hearts died out, very much as they would do with boron deficiency. A heavy application of pulverized limestone was applied broadcast over the plants. Four rows of plants were left without limestone for a check. The hearts of these plants and the older leaves made no further growth. Those that received the limestone recovered and made a normal growth. The grower told me he could see an improvement twenty-four hours after the limestone was applied. I have used this same treatment on spinach with equally good results. Even though this grower did grow some very good celery during the years we worked with him, it was necessary to apply some limestone every year to maintain a healthy growing crop. It seemed very difficult to kick the sodium out of the colloidal complex. My experience in later years convinced me that if we had applied 4 to 6 tons of calcium limestone per acre along with some gypsum, and had mixed it with the soil through the use of a rototiller set deep enough to break up the plow sole, we might have seen a permanent correction in two years. As it is, after some twenty years the grower is still having a problem, but it is easily corrected with a ton of limestone.

CHAPTER 4

Fundamental Research Must Be Given Preference

Let us never forget that the cultivation of the earth is the most important labor of man. Unstable is the future of a country which has lost its taste for agriculture. If there is one lesson of history that is unmistakable, it is that material strength lies very near the soil.
—Daniel Webster

Many of us are concerned about whether our grandchildren will eat. With proper planning and research, they will not go hungry; but we will have to reorganize our present thinking about crop production if we are to be assured of it. This is a serious question which concerns all of us, especially those who have the responsibility of directing the research programs which must be initiated to provide more food for the future. In discussing this problem we cannot be provincial. We must consider it in terms of our world resources. We still have agricultural frontiers with tremendous potentials for increasing our food supplies. Without delving too deeply into ways and means of growing more food, we probably should consider the means by which we can equitably distribute our food stocks. At the present time we have surpluses in isolated areas, while at the same time people in other areas are going hungry. This is a problem at which our politicians are nibbling. Until they decide whether it is more important to feed empty mouths or play at politics, there isn't much that can be done. Fundamental research and politics won't mix under the present scheme of operation.

So whether my grandchildren will have enough to eat will depend largely on how we approach the problem of providing food for future generations. I am sure we have the land. We have natural resources; but do we have the brains to know what to do with these natural potentialities? Can we control our rainfall? Can we control floods? They are all related. How well can we combine our efforts to assure ourselves as a world people of a supply of food for two thousand years hence?

THE PROBLEM

When we begin to speculate on our future food supply and the increase in the number of mouths we will have to feed, we are bound to think of all possible improvements that may be made. What are the avenues of research through which we can expect to increase our acre yields? I shall list them and discuss each in turn as they appear to me:

1. *Extending our frontiers into new land areas* still offers possibilities even though some people seem to think that there are no additional land areas. They assume that when yields reach a low level it costs too much to bring the soil back to worthwhile production. We are far from having exploited all arable land. Without further research, perhaps we have exhausted our good, fertile land; but with more fundamental research we can do much to bring "submarginal" land within the limits of good land. Of course we do have much world-wide opportunity to expand.

2. *Water supply offers some expansion.* Distribution of rainfall is extremely important. We may not know how to change that, but what we do with the rain that falls is of far greater importance. There are many areas at the present time where most of the rainfall runs off the land. As long as we have property-destroying floods we are very poor managers of the rainfall we

get. We not only wash away much of our topsoil but we lose many tons of plant food and lime, which we must replace at high prices.

3. *Dry weather and water supply.* In some areas one encounters both. Then there are areas where dry weather predominates through the whole growing season, but which have considerable rainfall during the off season. Holding that water in our soil is worth considering. We also have areas that have little or no rainfall, where irrigation is a must. In the first two cases we can do something without resorting to irrigation.

4. *Commercial fertilizers have far less effect* than most of us believe. We need much open-minded fundamental research, which must be integrated with our liming program.

5. *Temperature extremes have a controlling effect* on acre yields. There is again a very involved question as to what we can do about it, but plant breeding offers much in this field.

6. *Sunshine is an all-important factor* in our crop yields. Most people say, "Sure, it is important, but what can be done about it?" We can't do much about the weather, but our understanding of its importance can do much to help us modify other factors, such as the use of nitrogenous fertilizers. This field has been very much neglected.

7. *Soil types, conditions, and elevations* offer big opportunities for research and have potentialities for greatly increasing our world food supply.

8. *"Crop rotations" may be a misnomer.* Do they accomplish what most of us think they do? In many cases increasing the organic matter is too costly. We can increase it, incidental to other practices, if we decide that we need more in our soils.

9. *Organic matter in soils* may have good or bad effects on crop yields. Extensive root growth increases organic matter. We know too little about subjects 8 and 9.

10. *Can we agree on what is a productive soil* or are we confusing fertility with productivity? A highly fertile soil may not be a productive soil. What standards have we to judge yielding power? Too many people have the idea that in order to make a

soil fertile we must add manure and fertilizer. We must qualify our statements, because a mistake here can be very costly.

11. *Limestone makes soils fertile.* We have neglected our most important resource. We have a potential of a billion tons of food yearly if we learn how to make the best use of limestone.

12. *Cultural practices can change yields.* Subsoiling can promote bigger root systems. We must learn to farm many feet of subsoil instead of just the plowed layer.

The proper integration of all these factors, with all of them exerting a beneficial effect, can result in large yields per acre. Each particular soil type has a certain yield potential. We cannot expect the same yield on different types even when all our factors are exerting a favorable influence. And, of course, it would be expecting a great deal to expect the same influences to affect the crops in central Illinois that are affecting the crops in central Ohio, partly because weather and soil minerals will differ. We should not underestimate the varietal effect. I helped conduct a variety test in Virginia involving 72 varieties of corn, and the yield varied from 65 to 212 bushels. After three years the ten highest yields were distributed among practically the same varieties, but the varieties did not yield in the same order when they were grown in five different areas in the state. It is necessary to run these comparisons in each corn area to get the largest yielding variety for that area. It becomes a farmer's individual problem. However, from my own experience, I would say that if a person can't grow 100 bushels of corn without fertilizer, he had better investigate the management of his crop. Having sufficient calcium in the soil will practically guarantee the grower over 100 bushels, unless he does something radically wrong, like plowing the ground too wet or working it before the subsoil or A_2 horizon has dried out sufficiently so that it can be worked without puddling the clay.

Too many of our investigators have the mistaken idea that fertilizer alone will assure the grower good yields. That is the reason why our average yields are so low and have not increased materially since scientific agriculture was initiated.

Through my work in agricultural research institutions, I became aware of the shortcomings of our research staffs. Many of these people are in the wrong profession to do society much good.

Progress in research comes from original ideas. Very few men are capable of developing ideas. To accomplish something new is a gift from the gods. It would be considered fundamental research. The next step would be to prove or disprove a new idea. For everyone who has a new idea, there are too few who are capable of proving it. In other words, a hypothesis means little if nobody can prove it. A hypothesis is born in the mind of a gifted person. How long it takes to formulate the theory and prove it depends on the intelligence of people who work with it. Thus, one man can start research that will keep thousands of people busy for a lifetime.

When I first started in research work on crop production, I was convinced that our soils were woefully deficient in lime. Most of our soil calcium was solidified in tremendous layers where it was unavailable to our crops. I used the soil acidity tester for a number of years, but too often I was disappointed because I could not get better correlations between pH values and yields. I could get yield increases of 10 to 20 per cent, but some farmers by their unorthodox methods did a better job than I could do.

I next turned to soil tests, as a supplement to the pH tester, hoping to find the key to higher yields. I found out that many of our soils that had been heavily fertilized had a neutral pH but no available calcium. I published my findings in *Soil Science* in 1928, laying much of the blame for the high pH on the use of large quantities of sodium and potassium salts. Even though this had already been mentioned in the Russian literature, my paper was not well received by American crops people. My colleagues criticized my audacity in finding fault with the soil acidity tester. The people who were selling nitrate of soda and muriate of potash were very unsympathetic. I did not find fault with the acidity tester. There was nothing wrong with it. The fault was in the way it was used and the interpretation we placed on the readings. We were trying to test for something which could not be tested by such a method.

However, I soon found that testing for calcium paid off. Where two or more tons of limestone had been applied to a neutral soil having no available calcium, I increased yields 100 per cent or more. Now, some twenty years later, I still have arguments with agronomists hired by taxpayers as to the validity of the acidity test, and instead of bestirring themselves to initiate research and find out for themselves, they prefer to sit back in a swivel chair and say it isn't possible.

The use of high analysis fertilizers was an episode in my career which still rankles in my mind and for which few agronomists have an answer. With the introduction of 15–30–15 fertilizer in the late 20's, there was much speculation as to its value, even though the reduced freight rate could be a factor in shipping it long distances. I compared several of these with our standard 5–10–5, 5–8–7, and similar grades, on a unit for unit basis. The results were not good for the 13–26–13 and 15–30–15, and I assumed there was a reason why yields were lower with these high analysis materials. I compared a no-fertilizer control with my mixture and was surprised to find that the 5–10–5 decreased the yield slightly and the 13–26–13 decreased the yield even more.

These experiments ruined my complacency about the use of mixed fertilizers. I became bitterly critical and never published any results on fertilizer quantity experiments because I never had results showing any particular benefits from the use of mixed fertilizers. In later years, I did publish results on fertilizer placement studies, because I seemed to be working in the direction of better fertilizer utilization, until I found out that the better results I had with plowed-under applications were due to the fact that I had eliminated the root and seed injury where the fertilizer was applied in the row.

I continued working with high analysis mixtures because I wanted to find out why some farmers burned their crops with these fertilizers while others had excellent results. A 13–26–13 mixture was highly soluble and contained only ammonia, phosphoric acid, and potassium. A 5–10–5, slowly available, contained both nitrate and ammonia nitrogen, phosphoric acid as monocalcium and dicalcium phosphate, muriate of potash, calcium sulphate and traces of minor elements. About this time I chanced to talk with a greenhouse flower grower who was very happy with 13–26–13. I visited his greenhouse and found that he was fertilizing his plants in benches with what looked like sand. When I asked him what it was, he said it was 13–26–13 mixed with ground limestone. On further questioning I found he mixed one part of 13–26–13 with ten parts of limestone. His flowers were beautiful. I realized he had the answer to my problem. You needed the calcium for good results. I had underestimated the need for calcium. Also, he was using 13–26–13 much more sparingly than I had thought possible.

I must admit here that much of what I have learned about crop production did not come from textbooks. It came from what I observed in greenhouses and on farms managed by farmers, and I want to doff my hat to them. I am sure that many had no scientific explanation for their procedures, but they knew that they were getting results.

I continued working with high analysis mixtures, and found that because of their solubility, one could get as good results with a fraction of the pounds of plant food as one could with the low analysis mixtures. As a matter of fact, 150 pounds of 13–26–13 gave as good results as 1,500 pounds of 5–10–5. Also, one could get much bigger yield increases from the 13–26–13 if the soils were adequately limed. This information prompted me to apply the fertilizer in solution form, and I found out that the plants which I fertilized with 150 pounds of 5–10–5 in solution had just as much nitrogen, phosphoric acid, and potash in the tissue as those fertilized with 1,000 pounds of 5–10–5 in the dry form.

When I presented this information (very valuable to my understanding of the use of fertilizer) to the fertilizer dealers, they were not pleased nor did anybody applaud. Every dealer was seeing the volume of sales shrinking. In my enthusiasm to show how we could grow crops with less fertilizer at a lowering in cost per acre, I entirely overlooked the possible shrinkage of the bulge in the dealers' wallets, a very vulnerable spot, as I was to find out in later years. Needless to say, experimental data from high analysis fertilizers generally was not complimentary to low analysis fertilizer. The use of these materials never gained momentum in sales. It wasn't until the early 50's that the problem of fertilizer solutions again appeared, but it was recommended by our agricultural research agencies on a pound-for-pound basis. Reliable research results for or against the use of fertilizer solutions were not supported by fundamental research. To satisfy the dry fertilizer lobby, it is said that if you use 1,500 pounds of dry fertilizer per acre, you must use 1,500 pounds of solution. In the first case, you have a material that is not over 10 to 20 per cent available with a material that is 100 per cent available. You overfertilize with the solution because the plant gets five times as much fertilizer as it needs, and you produce succulent leaves and stalks and decrease the yield of grain by 50 per cent. One must produce dry matter.

The results I obtained from the use of high analysis fertilizers started me thinking along the lines of using fertilizer solutions. This venture almost wrecked my career, because the fertilizer industry did not like it, and my colleagues did not dare to agree with me. Beginning in 1930 and for twenty years, I was a lone wolf in the field. By making fertilizer solutions out of a 5–10–5 and comparing varying amounts of the solutions against the dry materials in equal amounts, I found that I could get maximum yields with about one-tenth of the fertilizer if I dissolved it in water. I was again walking where angels feared to tread.

Very little work has been done on the use of fertilizer solutions for crop production. Most agencies have the idea that if 500 pounds of 5–20–20 dry fertilizer was needed to grow a crop of corn, the same number of pounds of 5–20–20 solution was needed. There is no experimental evidence to support this statement. My experience is that you will reduce the yield of grain and fruit if you go beyond 50 pounds. In spite of all the years we have worked on fertilizers for crop production, we are woefully ignorant about their use, and there is no encouragement given to research organizations to find out much about it. My guess is that if there is any idea that such research will show that much less fertilizer will be needed if it is used in the solution form, there won't be any demand to conduct research on the use of fertilizer solution, unless farmers demand it because it will cut their cost of production.

I also had an idea that foliage sprays of fertilizer solutions might be valuable, but this practice is frowned upon by research people because so little fertilizer is expected to do so much good. The idea just isn't in keeping with our college training of future scientists. My guess is that our fertilizer industry is headed for some hard times if we continue along this type of thinking. Through its propaganda organization, the National Plant Food Institute, the fertilizer industry has set up a strong lobby to prevent any reduction in fertilizer usage. Because they have been successful in maintaining volume sales, they have maintained a fairly stable price for fertilizers. Should there be any reduction in sales of fertilizers, there will necessarily be appreciable increases in price. Therefore, even though a shortsighted policy is being followed by the N.P.F.I., it probably can be supported because it will be some years before our research organizations will dare to publish any data that would tend to reduce the acre cost of fertilizer. Even the farmers' own organization, the National Farm Bureau, is falling in line because they are in the business of selling fertilizer to make a profit.

Foliage sprays have shown their worth, but only a few agricultural scientists have conducted research on them. Most agronomists condemn their use even though they have no scientific proof of whether they have value or not. Yes! Foliage sprays have shown their worth. In the many comparisons I have made where I applied fertilizer solutions to the foliage of crop plants, I have had worthwhile yield increases slightly over 90 per cent of the time. In many of these cases the yield increases have been far beyond expectations based on the amount of plant food materials applied. This is a science in itself and is affected by as many factors as the use of soil-applied fertilizers.

"Ideas are born in the minds of men. Research is the tool that proves or disproves their validity." Many of those dealing with agricultural problems become confused in their thinking about the relationship and priority or sequence of ideas, hypotheses, theories and facts. A scientist has many ideas coursing through his brain. Some of these he can dismiss without further consideration. Some are worthy of being proven. A few keep gnawing at him, urging him to prove their validity. From such beginnings stems the progress made in our scientific approach to agriculture. Agriculture is only the application of scientific thinking to our way of life.

Life depends on food. Health depends on its quality. Without food none of us could exist. Our food supply almost wholly, directly or indirectly, comes from the soil. Our future as a nation or world power depends on how well we care for our soils.

In the earth on which we live, many chemical and physical influences have worked toward a common goal—that of providing a set of environmental conditions which has made it possible for the present human to develop and progress.

The elements needed to grow our food are all in the soil and are made by the tremendous floral deposits in the soil. We have large deposits of nitrogenous salts which can be mined and processed for application to our growing crops. We have large phosphate deposits which through man's ingenuity can be made available to our crops. We also have potassium and boron deposits. The one element which we probably need more than any other in order to grow crops successfully is calcium. Calcium, even though very abundant, is most often deficient. When this earth was formed, the existing chemical and physical conditions certainly provided mankind with everything he needed to supply his food needs for eons to come. It is up to our scientists to find out how to make all these minerals useful to man.

We have a tremendous supply of sea water, teeming with salts which are useful in the production of better food. And then along with this we have had a variety of climatic factors and forces which make it possible for us to make use of all the minerals in the soil.

Sunshine is our lifesaver, where moisture and temperature permit, because it gives us energy to keep our bodies in working condition. It provides us with starches and sugars, which in a 100-bushel corn crop form about 4,500 pounds of the weight of the saleable crop. The carbon comes from the carbon dioxide in the air, which is maintained by oxidation of organic matter (soil oxidation and burning of wood, coal and gasses), reduction of existing carbonates in the soil, and fermentation processes which exist because of optimum moisture and temperature levels.

Thus moisture, temperature (heat), sunshine, carbon dioxide, and mineral elements in the soil all add up to food for humans and animals. Only the proper contribution of each of these can result in the maximum yield of high-quality food. Optimum conditions could boost the average yield of corn from 63 to 120 bushels or more, the average yield of wheat from 27 to 60 bushels and even 100 bushels, with irrigation, and all other crops in proportion. We have much land in the United States that potentially is 100-bushel corn land, which now is not producing 30 bushels of corn or 10 bushels of wheat, which with the expenditure of less than ten dollars would grow 100 bushels of corn. The experiment reported in Table 1 was conducted on a 10-acre field that would not grow 50 bushels of corn. It had been in corn

TABLE 1

EFFECT OF LIMESTONE ON YIELD OF CORN ON A BENNINGTON MARENGO SOIL

Treatment	*Bushels No. 2 Shelled Corn per Acre*
1. No fertilizer, no limestone (6 tons calcium limestone per acre needed)	43
2. 300 lbs. calcium limestone applied over the row	60
3. 2 gal. Growers 10-20-10 Fertilizer Solution applied on seed at planting	54
4. Combination of 2 and 3 5 replications per plot	103

the previous year. This experiment has been repeated several times with similar results and I should like to encourage any farmer who has not tried this little experiment to do so. The fertilizer solution was applied directly to the seed with a special attachment.

Most of our results showing spectacular increases in yield have occurred on soils which have been classified as submarginal. It is much more difficult to double the yield on supposedly good land where an abundance of fertility exists. I have planted corn on heavily fertilized land which the grower admitted would not grow 65 bushels of corn, applied the necessary calcium, and found that my method did not show response until after the third year. Fertilization with nitrogen can account for this, and yet my critics have stated the reason I got worthwhile results was because there was so much fertility in the soil from previous years. This was easily proven to be a false premise. With my program, yields increased in later years as this fertilizer was removed through crop production.

Whether a soil needs limestone and, if so, in what quantity, may be discovered through an experiment such as was carried out on a farm in Washington C. H., Ohio (Table 2). I checked a field which would not grow over 65 bushels of corn and would not grow clover or alfalfa. The soil acidity test showed no need for lime, but the calcium test showed a deficiency of calcium equivalent to 7 tons of high calcium limestone. This soil was

TABLE 2

EFFECT OF FINELY GROUND LIMESTONE ON YIELD OF CORN ON A MIAMI CELINA SILT LOAM HAVING pH OF 7.1

Treatment	*Bushels No. 2 Shelled Corn Per Acre*
1. No limestone	63 ± 3
2. 2 tons plowed under	65 ± 3
3. 4 tons plowed under	65 ± 3
4. 6 tons plowed under	65 ± 3
5. 8 tons plowed under	123 ± 4

potentially fertile, but needed many tons of limestone to supply the calcium needs to a depth of three feet. Even though the soil acidity reading was satisfactory, it took 8 tons of limestone recommended to increase the yield. Had we not applied more than 6 tons, we could easily have assumed that the soil did not need limestone.

HUMAN HEALTH MAY DEMAND CALCIUM

I have received numerous letters from women asking me to give my support against the use of commercial fertilizer for growing vegetables. Of course, if chemical fertilizers are harmful to humans, then they are harmful to animals, and we should not use them. Some people have become imbued with the idea that commercial fertilizers are responsible for malignant diseases and for lowering our general health. If this were true, we should have had a definite increase in the occurrence of malignant diseases since the use of commercial fertilizers first began. I doubt that this can be proven. I do believe that the deficiency of available calcium in our soils could contribute materially to undermining the health of humans in this country, not because the fertilizer is there but because something else which humans need is missing.

In other words, we must satisfy the lime requirement of our soils first. Then add sufficient plant nutrient materials (fertilizer ingredients) to prevent deficiencies of phosphorus and potash from occurring. This does not answer the critics who expect us to grow good crops without commercial fertilizer, because we must consider the loss of fertilizer from erosion and leaching, besides what plants remove. Organic gardening people assume that the phosphorus that comes from rocks is different from that which comes from manure. Animals live on grass which is grown with the phosphorus, which, because of weathering, becomes available from the complex minerals in the soil. The only difference is that the minerals used in commercial fertilizers are treated with acid to change the phosphate rock to soluble phosphates—but this is just what weathering in the soil does, only faster. People seem to be afraid of the word "acid," and yet every process in the soil as well as in the growth of the human body and in plants involves the action of an acid on an alkali at some point between the time the minerals are absorbed into plants and animals, and the final end product. Concentrated acids may be very corrosive, but in their diluted forms they make it possible for humans to exist on this earth.

The controversy of organic versus commercial fertilizer farming is not sufficiently well understood to warrant making an issue of it, as some overzealous individuals would do. If we understood our soil processes better, we could use both organic and inorganic plant foods to advantage. Furthermore, from what little I know of the use of commercial fertilizer in the production of crops, I see no reason why we have to take sides. You can honestly pick out advantages and disadvantages on both sides, because we usually don't do enough research to improve our products. However, if we prepare the soil so that we recover the most fertilizer value from organic fertilizers, we will also get the most efficient returns from commercial fertilizers.

Let's decide, first of all, what we want. We want food crops that are grown on soils that support large yields, because we know that those crops are well supplied with all the necessary minerals and vitamins needed by humans to build healthy bodies. And, to go further, we can assume that all the minerals needed to grow a good healthy plant are also needed to grow a healthy human being. If we have made a mistake, it is because we have oversold commercial fertilizer and undersold liming materials.

If the yield of a crop is low, it means that one or more of the minerals are missing. That means that food produced on that deficient soil is not good for humans because it is lacking in some mineral needed by humans. If I were to criticize our commercial fertilizers, I would criticize the way in which humans use commercial fertilizer rather than the ingredients in the fertilizer. If there is anything wrong with our food, it is because of something that is lacking and not because of having some phosphorus or potash or magnesium or any other mineral element. Perhaps we could improve the fertilizer mixture through the choice of ingredients. Even too much nitrogen in our food plants (unassimilated nitrogen—nitrates, ammonia, soluble amino acids) is usually the result of imbalance. It may lower the food value; but when we learn more about it, it probably will be attributed to a lack of calcium or phosphorus.

For some reason people pick on phosphorus. Our dealers called it acid phosphate at one time. Then it was superphosphate. Some seem to think that organic phosphates are less harmful than the mineral phosphates. And yet those same people take dicalcium phosphate pills and think nothing of it. They drink a lot of phosphoric acid in their soft drinks. I would prefer the mineral phosphates purely for economic reasons. It keeps our costs down.

Some people like to think that cancer is caused by commercial fertilizer. It would seem more reasonable to assume that it is caused by some deficiency—a lack of some of the minerals contained in fertilizer or even a deficiency of calcium in the soil. Besides, if we were to stop using commercial fertilizer, crop production could eventually drop to a point where we could not feed the people in this country. Our research men have the responsibility of finding means of increasing yields with its use. Up to date, they haven't done a very good job. We must produce more meat per acre.

NITRATES AND WELL WATER

The presence in well water of nitrates above the threshold of toxicity has prompted trained men to investigate the possibility that this excessive nitrate is poisonous to animals and humans. In some areas in heavy soil, wells were condemned for drinking water. Veterinarians have attributed the sudden death of animals to this "vicious" killer.

The heavy application of nitrogen to our soils is blamed for this, but I don't believe that the facts can be proven. It is difficult to tell the nitrate from the water (although the method used for removing salts from sea water with electricity could be used for this purpose). Nobody has any proof that the nitrate wasn't present long before any nitrogenous fertilizer was applied to the soil. Perhaps it always was present but never came to the attention of chemists.

The people who recommend commercial fertilizer for crop growing are human and feel just as affectionate toward their children and loved ones as those who condemn the use of it. Because they do not have the scientific facts to back up their recommendations they know that they can safely eat foods grown with commercial fertilizer. The people who belong to the "organic cult" may not be trained scientifically and may not have the scientific data to back up their statements. As a result, they are careless with the truth. Scientific facts, since the time of Galileo, have not met with popular response from the public even though the welfare of the human race is becoming more and more dependent on scientific discovery. Scientifically trained nutritionists are ferreting out the facts so that we may better understand ways and means of improving our health and welfare.

ORGANIC MATTER HAS MANY USES

The questions of the importance of organic matter, its function, and its value as a source of plant food, are not well understood even by some of our soil scientists. For this reason the preference of organic farming over the use of commercial fertilizer can be argued with little fear of contradiction. It fits in more closely with natural phenomena and may be quite true, but to date we still need more proof. We all appreciate that a certain amount of organic matter in the soil is a good thing. It helps to mellow the soil when calcium is present. It supplies some plant food and it does help to buffer the soil. It all adds up to better crops; but, to say it is the ideal, is begging the question. Too much organic matter makes the soil black where poor drainage exists. Under such conditions, it could cause damage to crops because of interference with the oxygen supply.

Organic plant food is often considered superior to chemical

plant food. On sandy, well-limed soils, organic fertilizer may produce better crops than commercial fertilizer, not because of any superiority but because of differences in availability. It is not difficult to retard growth with too much salt concentration. This, however, does not mean we should increase organic matter. Better plants have bigger roots and leave more organic matter. Farmers who grow crops to make a living cannot afford to build up organic matter in their soil unless they can be assured of yield increases sufficient to pay the costs. The amount of organic matter is determined by climate and intensity of cultivation. When a heavy crop of green stuff is plowed under, it has very little effect on the organic matter, but it does add a large amount of organic plant food, which the next growing crop soon uses up. Organic matter is built up from fibrous material found in the roots. Thus, the more roots that can be grown, the more organic matter will remain in the soil.

The minerals and salts in a fertilizer bag are the same as those found in any plant and are the same as those found in animal bodies. There is no reason for the public to assume that there is anything wrong with them. Phosphorus is combined with calcium, both of which are very essential for good human health. Potassium, tied up with chloride or sulphate, is essential to humans (in not too large quantities) as well as to plants. Nitrogen, either as nitrate or ammonia, when taken into the plant is soon changed to proteins. The calcium and magnesium come from limestone, with which many of our soils are not well supplied. On this basis, I can see no reason for all the criticism of the use of commercial fertilizer—except in our ways of using them. Too much of many things can be harmful. My only criticism is that we haven't learned how to use it to the best advantage.

There are mineral elements which can be applied to the soil or are released in the soil which plants will absorb and store in their tissues and which are toxic to animals and plants in rather low concentrations. However, since they are not included in our fertilizer mixtures, there is no need to be concerned about them. Adequate calcium in the soil will prevent any toxicity that could be associated with the use of mineral elements, and thus the importance of providing sufficient quantities of limestone for our soils takes on more significance.

If I were to pick any one thing out of the fertilizer industry which might raise a question in my mind, it would be the application of greater amounts than are actually needed. Most crops don't suffer from a lack of fertilizer so much as they do from an imbalance of nutrient materials.

CHAPTER 5

Crop Yields, Plant Food Materials and Our Natural Resources

PROFITABLE CROP PLANTS are grown on many different soils containing wide variations in nutrient content, and in areas of widely different climatic conditions. Climatic conditions probably have a greater effect on maximum yields than the potential nutrients that soils contain. All soils contain large amounts of minerals which, as a result of the proper weathering, are capable of producing satisfactory yields. The problem the research man has is to find out, first, what the key variable is that controls the productivity of the soil and, secondly, what must be done to release the plant food material so that the plant can absorb nutrients and use them efficiently for optimum growth. Also, since there is much available plant food lost through erosion, means must be found to steer this plant food into the plant instead of permitting it to move to the rivers and the oceans. Many of the bottom lands or flood plains of streams are the result of erosion (but, for unknown reasons, are not producing good yields).

The word "productive" is prominently used in discussions of land use and soil classification. Whether or not soils are capable of producing large yields depends largely on available nutrients rather than on total potential nutrients. Even though soils are well supplied with available nutrients, there is no guarantee that the soil will be productive.

We have a group of soils which are classed as submarginal—

those which for one of many reasons do not produce profitable crops. My experiences with many of these soils have given me a different approach to their use. In most cases, these soils are lacking in calcium. If they are too rough to farm, they can have limestone applied, a ton or two per acre at frequent intervals, and be made to produce excellent pasture. Some can be terraced and strip-farmed and properly limed, and others, which are flat and which produce only poverty grass, can be heavily limed to make them highly productive. Calcium deficiency is usually the controlling factor. There are very few farmable submarginal lands which cannot be made highly productive with the application of sufficient limestone.

There are also many potentially highly productive soils which do not make any money because the farmer tries to correct something that is not at fault. He throws on more fertilizer when actually he has too much on. Such soils have responded to limestone applications.

Clay knobs and gravelly ridges give farmers headaches because they usually have a high pH and still won't respond to fertilizer applications. On occasion, they have been referred to as alkali spots, when actually they are deficient in calcium.

I worked with a farmer near Radnor, Ohio, who had three barren gravel ridges traversing one of his 35-acre fields. These had grown practically no crops for many years, particularly during hot, dry summers. Six tons of limestone was applied over a period of two years. Two years in succession this field averaged over 135 bushels of No. 2 corn with the tops of the ridges producing just as good a yield as the slopes and bottom lands. The fodder was not as tall nor as heavy on the tops of the ridges. Even though this is limestone soil, calcium availability was too low.

A farmer in southern New Jersey asked me to work with him on a farm where each field had at least one pond hole. The water in some of these ponds disappeared during the summer. Several did not dry up even in dry weather. We checked the soil for calcium and applied 2 tons of pulverized limestone where the ground was dry. Then we subsoiled the fields 21 inches deep in the vicinity of the ponds by circling them at three-foot intervals while applying some pulverized limestone (600 pounds per acre) in the subsoiled trench. The ponds gradually dried, apparently because we prevented water from seeping into the depressions. When the depressions dried, we subsoiled them and applied 2 to 4 tons of pulverized limestone and proceeded to crop the area. Potatoes, corn, and tomatoes gave good yields in all these areas and water did not accumulate in later years.

A farmer in southeast Ohio bought land which was classified as unsuitable for crop production and proceeded to farm it. He was told it was submarginal. Somebody had told him to apply limestone, which he did. He told me that he had applied 6 tons of limestone per acre and had grown over 100 bushels of corn per acre for six years; and he had not applied any fertilizer to this soil. He said he was getting better yields than some friends who were applying over 600 pounds of mixed fertilizer on what was considered very fertile soil. This proves my contention that we don't use our heads, even though crop production depends on how we interpret observations. We *can* use too much fertilizer when it is *not* the factor controlling a good yield. There is something wrong in our thinking and in our approach to research work. We don't consider the factor that actually controls yield. We have given credit to the wrong practices.

Every year I am surprised when I talk with farmers on the black prairie soils of Illinois. They tell me that they are lucky to get 100 bushels of corn per acre on land that has enough fertility to grow over 300 bushels. There must be many who don't get 75 bushels, if the average yield in Illinois means anything. I tell them they should see some of the soils on which our Ohio farmers are growing 135 bushels an acre. They naturally are very skeptical when I tell them it is a gravelly loam soil. Even in Ohio, what we consider our best soils do not produce our highest yields.

I have had many cases where farmers living on deep, fertile limestone land have attempted to grow 150 bushels of corn by following "sure-fire methods." The sure-fire procedure is to apply 400 to 1,000 pounds of mixed fertilizer before plowing a clover sod previously covered with manure. Then they plant the corn with a starter of 300 to 600 pounds of fertilizer and side-dress with 100 to 150 pounds of nitrogen. With a fair distribution of rainfall, the corn grows rapidly. The stalks are tall and heavy and the field produces a big tonnage of fodder for silage. There may be approximately 13,000 stalks per acre if germination is good. There are no ear stalks because the foliage is dark green and succulent.

The grower follows this practice to grow 150 bushels of No. 2 corn in contradiction of my recommendation. I applied 2 tons of pulverized limestone per acre and only 2 gallons—22 pounds—of 10–20–10 fertilizer solution in the row on the seed for a starter. The foliage is sprayed with 2 gallons of 10–20–10 fertilizer solution in the middle of July. I had 14,780 stalks per acre; apparently there was less interference with germination.

When these plots were harvested, the plot with fertilizer solution yielded 135 bushels on 7- to 8-foot stalks; whereas the farmer's plot yielded 59 bushels. The average farmer, viewing this field on August 1, would have guessed that the heavily fertilized plot would have outyielded my plot by at least three to one, because the foliage was dark green and voluminous.

What was wrong and what is wrong with our thinking? Why do we make such fertilizer recommendations? What kind of data are these recommendations based on? The practice outlined is considered necessary for a big yield, but a big yield of leaves and stalks does not make good silage.

We examined the field carefully. On the dry fertilizer plot, the corn was 12 to 15 feet tall, compared with 7 to 8 on the solution plot. Thirty-nine out of 100 stalks were barren (no ears on the stalks) and only 9 out of 100 stalks had ears over 8 inches long. The remainder of the stalks had nubbins. On the fertilizer solution plot, 97 out of 100 stalks had ears 8 or more inches long.

MINERALS FOUND IN PLANTS DO NOT INDICATE FERTILIZER NEEDS

We have many farm advisers and people in the fertilizer industry who say you must apply 2 pounds of nitrogen, 1½ pounds of phosphorus, and 2 pounds of potash to produce a bushel of corn, and that it is a simple matter to figure fertilizer needs. If you grow 50 bushels and you want 150 bushels, all you need to do is to apply 200 pounds of each of the fertilizer ingredients. They ignore the nutrients supplied by the soil. This may be enough to grow the crop. Why apply more? I know that if you analyze the leaves, stems, and grain or fruit on a crop, you will come up with a figure showing certain amounts of minerals. This does not mean that you need to apply that much. Very seldom are check plots left for comparison and, if they are, they are forgotten at harvest time.

You will find that 95 per cent of the plant comes from water, carbon dioxide, and sunlight. The minerals in the tissue which we supply as fertilizer are very few indeed. The nitrogen helps to keep the plant green and builds up proteins with the help of starches and sugars. But the starches made in the leaves and stored in the seed make up the bulk of our yield of grain. Phosphorus is very important in these processes. Potash acts as a policeman to make these chemical processes take place. It is not known how much is actually needed. This probably depends on many factors. Any good soil, properly handled, will supply an abundance of these plant food materials.

Much data has been collected showing the minerals present in crops grown in various parts of the country. They vary tremendously. One hundred bushels of corn grown in one area may contain 12 pounds of phosphorus, while in another it may contain 60 pounds. The important consideration is that in both cases 100 bushels of corn were produced. At the present, we must assume that much of the data was used for sales purposes, and probably had little basis in fact. It is true that you can show some increases in yield with some additional phosphorus or potash or nitrogen, but when it means an additional 10-bushel yield with no additional profit to the grower, it does not lend itself to any extreme enthusiasm, particularly when the yield is only 75 bushels per acre, or a comparable yield in any other crop.

There is an abundance of raw materials available for processing and making mixed fertilizers, and we have a tremendous capacity to manufacture mixed fertilizers from them. Since they can make it, any concern doing this type of manufacturing must sell it, because it is interested in making profit. A high-powered sales manager, who probably doesn't know anything about a farm or conditions on a farm, proceeds to load the farmer with plant food materials, many of which he may not need. If he needs it, he may add to his profits. If he doesn't need it, he is burdened with an additional tax because of the cost of the fertilizer.

So we take the path of least resistance and, using the formula, proceed to sell the farmer fertilizer whether he needs it or not, in quantities that should give him 50 or more bushels above his present yield of 60 bushels. If the yield increase is sufficient to make a profit over the additional cost, everyone is happy; but the farmer still has a bigger fertilizer bill. The fact that he may need only a few pounds of phosphorus doesn't enter the picture. Generally, it is a good bet that he has bought something that he doesn't need; but as long as he doesn't run comparative plots with and without the fertilizer, he won't know—unless he counts the money left in his pocket after he pays his bills. If he doesn't get an increase in yield, he is told he will get it in the crop the next year. This is apt to be wishful thinking. I have taken records on farms for ten years after 1,000 pounds of fertilizer was plowed under, and after ten years the grower was still trying to find some return in yield for his fertilizer.

Given the presence of sufficient calcium in the soil and an adequate supply of plant food materials, climate and agricultural practices will produce a certain given yield. And if that crop is analyzed, the mineral content will reflect the amount in the soil. If there is an abundance, the tissue will show a higher level. We can be sure that if we produce a 100-bushel yield, there will be present in the soil and in the plant tissue nutrients sufficient to do the job. We have no information to say what percentage of mineral elements is necessary to produce 100 bushels of corn. This depends on too many factors, the important one being the amount of oxygen bathing the roots of the crop. Foul gases around the roots do not produce good yields. Adequate calcium in the soil will detoxicate the gases formed by fermentation.

The sensible procedure is to set up a series of plots to find out whether there is a shortage of any mineral. The sales pitch is that there never is enough in the soil—farmers must be sold what they can afford to buy.

The distribution of rainfall has a lot to do with the growth of crops. We expect a normal distribution, and the resulting crop will be good if temperatures are normal. In 1957, the corn crop in Ohio had too much water at planting time. Corn did not germinate well because the soil, in many cases, ran together and became puddled on the surface. Oxygen was excluded from the seed. This excluded oxygen from the roots. The plants turned yellow, and many people attributed the yellow color to a lack of nitrogen. Since nitrogen is assimilated in the fine feeding roots, nitrogen could not be used by the plant, because the roots were suffocated. After the soil dried and oxygen was available, the plants made new roots, after which the leaves again turned green; but the corn had been stunted and never recovered sufficiently to produce even a fair yield. Fertilizer had little effect.

In 1958 the heavy rains came a month later and similar conditions developed. Corn and other crops that had a good start were not hurt as much. Crops on soils well supplied with calcium fared much better than on those where lime was needed. The corn again turned yellow, and again nitrogen deficiency was blamed. Those that applied nitrogen probably did more damage than good, because there were no feeding roots to utilize the nitrogen. Foliage sprays with fertilizer solutions were very effective in starting activity in the plants, and caused the foliage to turn green. If applied at once, this prevented severe stunting of the plants. It was thought by many that the fertilizer had leached away. The surplus water did help the dry fertilizer, because more of it became available to the plants in a short time. This hurt germination of seed and produced some poor stands of corn, which reduced yields. In such cases, where no fertilizer had been applied there was a better yield than where appreciable quantities had been applied. Some growers pulled a 16-inch-deep subsoiler between the rows and added 30 to 50 bushels of corn to their 50-bushel yield.

From my experience with comparative plots with and without additional fertilizer (probably because of so many interfering factors), after seven years the plots without fertilizer produced better yields. We have been conducting these farm experiments for twelve years, and this is the main reason why I have much more faith in my method of fertilization than in that recommended by our farm agencies. If it were true that we could use a (N—P—K) formula to increase our yields, average yields would be more than double the present averages and, if we were building up fertilizer reserves (carry-over of more fertile soils), our yields should be much, much higher. This suggests to me that we may not be aware of the real situation and that what we refer to as worn-out soils are not created by insufficient fertilizer.

Our weakness is that we believe this propaganda, that this formula is the true yardstick. (I call this "propaganda" because I haven't seen or been able to demonstrate its validity.) Illinois' highly fertile prairie soils can't grow 100 bushels, even though it has sufficient plant food materials made available every year to grow several crops of corn, while poor soils in Ohio, with adequate calcium, can grow 200 bushels. Nor can growers in Ohio demonstrate on low-fertility fields that this formula holds true, because insufficient available calcium is the limiting factor.

When I talked to a group of farmers in one of our hill counties, I was informed that an authority had told them to double the amount of fertilizer if they wanted to double their yield. One grower said that he had tried that and it had reduced his yield. My answer was, "Perhaps you don't need fertilizer." The next year I convinced this grower to put on four tons of limestone per acre. His yield increased by 50 bushels—without adding any fertilizer.

Plants need certain things to produce a crop. We know that if the plants can't get them from the soil, they must be applied. We also know that there is a lot of plant food material made available every year through the soil's chemical processes (weathering and oxidation), which are supported by the application of limestone. How much is made available depends on weather conditions, potential minerals, organic matter, type of soil, and numerous other factors. We try to test the soil, but so far our tests are about as good as a broad guess. I use these tests for calcium and phosphorus and, even though I can double the yield of corn by following the readings, they are still far from accurate. I feel that if I did not have thirty years of experience to go along with these tests, they probably would be of little help.

Farmers have asked me why they can't get as good yields as their neighbors when they follow the same practices. The answer, necessarily, is that they don't have the same conditions, and actually don't follow the same practices, even though they think they do. One of my customers planted corn on May 9, and because of too much water (rain), had to replant half of the field on May 28. Except for working the ground, which probably speeded up oxidation, the program was the same. The May 9 planting yielded 63 bushels; whereas the May 28 planting yielded 196 bushels—with no additional fertilizer.

If we are dealing with an annual plant (planted and harvested the same year), we must attribute more importance to small details than if we had planted tomatoes, which would grow indefinitely if frost did not kill the vines. The corn planted May 9 was checked temporarily, because the heavy rain packed the soil, even though the germination was not hurt. Loosening the soil on the May 28 planting and temperature set the stage for rapid growth after the seed germinated. The soil was not packed again by heavy rain.

This occurred in 1956. I am sure that the same results would not have been obtained in 1957, because the weather sequence was so different. Time-of-planting experiments from one year, therefore, mean very little to the grower, because results will be different every year.

When we talk about fertility needs, which interest salesmen the most, we are probably talking about a detail in production which is relatively unimportant. I know that many will disagree with me on this, and from the standpoint of one who has something to sell, I wish this were not true, but I have to admit it whether I want to or not. If I were talking about the calcium requirement of the soil and the importance of a proper calcium saturation of the base exchange complex, I could not make this statement. It seems to me to be the key to our whole fertility problem. I probably place more importance on it than on any other detail. But even this was not true in the case of the corn planted on May 9 and May 28. Here climatic and soil compaction certainly played the major roles. I watched the corn grow at frequent intervals, and at no time during the season was there any slowing down of the development of the stalk and the ear on the field planted on May 28.

CHAPTER 6

The Soils That We Farm

Soils vary in their composition with respect to total available plant food, organic matter, clay, sand, silt, and lime content. They also vary with respect to drainage, aeration, and natural moisture content. All these factors are affected by temperature, rainfall, cultural practices, and the crops grown on them. Topography has a tremendous effect on the productivity of any soil.

Because any one factor can have an effect on a soil, it means that all other factors are affected as changes occur in the prevailing temperatures, the prevailing rainfall, or the lime content. Even the addition of 1,000 pounds of potash, in addition to supplying potassium to the crop, indirectly affects all other changeable factors in that particular soil. For one thing, it can release calcium in a soil by replacement, which could account for a good yield increase. So, when we deal with means of improving the productivity of a given soil, we have to do a lot of guessing, in spite of the fact that we supposedly have good soil tests with which to determine the nitrogen, phosphorus, potash, calcium, magnesium, iron, and aluminum, in addition to the reaction (pH, sour or sweet condition).

To the scientist, soil is a mixture of materials of different sizes. To the farmer, soil is what he grows crops on. If he has a level, black soil, he can't understand how a neighbor can grow crops on a grayish-brown, hilly soil. Yet, we see high yields produced on every kind of soil, from the thinnest sand to the heaviest clay soil. When we refer to the soil as light or heavy, we do not refer to weight. A coarse, sandy soil, which we call a light soil, actually weighs more per cubic foot than a clay soil, which we call heavy.

Sandy soils are easy to work with farm tools, but they have a minimum of plant nutrients. We refer to them as light soils. A clay soil, which has a high percentage of clay in it, is a potentially fertile soil, but not easy to work, so we call it a heavy soil. But when it comes to growing crops, it is possible to harvest as good a crop on one as on the other.

If you were to place pure white sand in a bench or tub and supply a nutrient solution to it, you could grow a good big ear of corn, just as you could in a "good" soil. A pure sand must be supplied with nutrients and water at least every other day. It has no capacity to hold nutrients. It has practically no base exchange. In other words, some nitrogen, phosphoric acid, potash, sulphur, calcium, magnesium, manganese, boron, iron, copper, zinc, and a few others in very minute quantities, with sufficient fresh water to keep the plants from wilting, will grow a good stalk of corn with a good ear on it—in pure sand. Of course, it is essential that the plant be grown in full sunlight at temperatures between 50 and 90 degrees. If you omit sunlight, you can't grow anything. Sunshine and temperatures are the controlling factors. Without them and carbon dioxide, we can't produce starch and proteins in the plant.

Suppose we take a sand such as we have in some areas near bodies of water, and grow plants in it. We can grow a fairly good plant with just water, because the sand has some nutrients in it. To grow a good plant on sand we must add some nutrients, but many fewer than in a pure sand culture. It must be supplied with nutrients at semimonthly intervals. It has a trace of clay and organic matter, which gives it a small base exchange capacity, and for this reason it must have 500 to 1,000 pounds of pulverized limestone per acre added to it, to build up the calcium saturation. Without this, you probably would not grow much on it, in spite of plenty of other nutrients. A half-inch of rainfall can do a lot of good to a crop growing on these soils.

A loamy sand has a little more clay and organic matter in it. It has a higher base exchange capacity and therefore needs more limestone to neutralize its negative charges. These soils need 1,000 to 1,600 pounds of pulverized limestone per acre-foot (6⅔ inches) to supply the calcium to approach 85 per cent saturation, which seems to be necessary to grow a good crop. This soil has more potential fertility and therefore need not be fertilized as often as a sand. Animal manures produce a good crop because the soil gets well supplied with oxygen, and, if adequate water is available, it supports good crop growth. Because of the higher base exchange capacity, it will hold plant nutrients longer, and crops usually don't need to have fertilizer applied more than once or twice during the growing season. As soils become heavier, we have stronger buffer systems, which must be reckoned with when we apply lime and fertilizer.

Sandy loams are among our best soils. They have a small amount of clay and 0.5 to 1.5 per cent organic matter in the surface foot. In the North, they may have 2 per cent more organic matter than they do in the southern United States. Since the clays and some organic matter are chemically active and contain negative charges which readily combine with basic materials, sufficient limestone should be applied to furnish the calcium and magnesium to neutralize the negative charges. Eighty-five per cent of those charges must be neutralized with calcium to make it possible to grow a large-yielding crop. Any saturation less than 85 per cent makes it more difficult for crop plants to get sufficient nutrients.

The more base exchange capacity a soil has, the more potential fertility it has. Sandy loam soils are usually quite fertile and, if properly processed, will produce as good a crop as can be grown under our yearly climatic conditions. They will hold a fair amount of calcium, which will last five to ten years. These soils are usually well aerated where drainage is good and, when properly limed, they permit water and nutrients to move readily from the surface to the subsoil and vice versa. On the other hand, they can also become troublesome if they are not adequately limed.

Plow soles, originally of geologic origin, form at the bottom of the furrow and are aggravated by our cultural practices. If the soil becomes devoid of calcium and is plowed at the same depth every year, these plow soles build up and become troublesome. They become very hard when dry and are very sticky when wet, because the clay which has accumulated becomes hydrated as the calcium is removed during the years of cultivation. Such plow soles prevent the roots of crops from penetrating into the subsoil, and thus can curtail crop growth in dry weather. The plow sole may prevent moisture from moving up from the lower depths.

Fertilizer, when properly applied, need only be applied once during the season in an adequately limed sandy loam. These sandy loams are underlaid with a subsoil that contains appreciably more clay and silt than the surface soil. The calcium requirements of a sandy loam soil may be between 1,600 and 2,400 pounds to an acre-foot. In a non-limestone soil, this may be multiplied by four or more, depending on how deep the roots will readily penetrate. If there are no limestone minerals in the subsoil, the application of 10 or more tons of pulverized limestone may be required to get maximum yields even in years when temperature and rainfall are ideal.

Sands to sandy loam soils are easy to work. They are not readily puddled if plowed slightly wet. They do not become sticky like clay soils, and dry out soon after a rain. They should not be worked into too fine a seed bed in areas where heavy rains have a chance to pack the surface after a crop has been planted. Too much preparation of the seed bed will dry them out. This prevents the seedling from making a good root system. Many farmers are learning to plow later and plant immediately, in some cases delaying planting until after the mid-spring rains are over.

Before we go any further in this discussion, we might stop and point out what makes up a surface soil. Each soil type has different characteristics. Stones, gravel and other mineral additions are local ingredients. See Table 4 (page 156).

Clay and organic matter are the only two ingredients which affect our discussions of fertility. All soils have different minerals and vary in composition. In general, organic matter is high in cold temperatures and low in hot climates. Thus, the calcium requirement will vary. In general, a pound of active organic matter has four times as many negative charges and, therefore, requires four times as much limestone as a pound of clay. These figures are rather inexact and are difficult to compare. However, on the basis of field results, high organic mineral soils need much more limestone than clay loams. I have seen greenhouse experiments on high organic matter soils which required 100 tons of limestone per acre to make up a soil to place in 5-inch beds to get worthwhile results.

Silt loam soils are heavier and require more care in timing cultural practices. They can be puddled by working when too wet. They contain more clay, although the organic matter may not be much higher than in the sandy loams. Manure should be used only where drainage is good and organic matter is low. The lime requirement of the silt loam is only slightly higher than that of a sandy loam. The base exchange capacity is equivalent to 2,800 pounds, more or less, of calcium. They are well-buffered, productive soils. In acid soil areas, their lime requirement may run up to 16 tons or more per acre, because there is so little calcium in the subsoil.

In limestone areas these soils can be limed with 4 to 6 tons in the surface plow layer. There is usually plenty of limestone in the subsoil. However, it is always good insurance to test the subsoil, because often it is glaciated and varies greatly on the same farm. I tested soil in an orchard where one row of trees was growing fine because there was plenty of limestone in the subsoil, while the adjacent row was on a low calcium soil. The rocks had been folded when formed and the limestone stratum was at the surface under one row and down 50 feet on the next row. It is usually a good idea in rolling areas to take samples for testing on the hills and in the hollows. One often finds that because the organic matter is higher in the valleys and because there has been surface erosion, the actual calcium available to the crop is much higher in the valley.

We have a lot of crop land that can be classified as silt loam which, when it is adequately limed, will produce some of our larger yields of corn, along with our sandy loams. This is sticky when wet and tends to bake when dry, but if a person will go fishing when it is just a little too wet to plow, he can grow 200 bushels of corn or 800 bushels of potatoes with little difficulty.

Clay loams are difficult soils to farm. When adequately limed they are easy to cultivate, plow, and harrow, but with insufficient lime they cost the farmer money. They are potentially very fertile but very often show no response when fertilizer is applied. When limed they are easy to farm, providing that the farmer is not in too much of a hurry to get his crop planted. There are clay loams along our eastern seaboard which, when sufficiently limed, can produce 300 bushels of corn without the addition of fertilizer.

I worked with a grower who had some Lenore clay loam which in forty years had never grown a crop. We applied 6 tons of limestone per acre and disced it into the surface to a depth of 10 inches. Then we applied 2 tons per acre with a subsoiler which penetrated 2 feet every 36 inches, first in one direction and then at right angles. Corn was planted that year, producing a tremendous crop. People came from fifty miles away to see it. The grower told me he sold 98 tons from 20 acres, which was close to 175 bushels an acre. After that experience I became very much interested in rehabilitating our clay soils. Because these soils are so dense and limestone penetrates slowly, it is necessary to use a subsoiler to make the limestone penetrate faster. The subsoiler should be used when the subsoil is dry, because it tends to break up the subsoil in all directions and gives the limestone a chance to penetrate into the cracks wherever there is sufficient rain to wash the surface soil down into the cracks in the subsoil.

I have had many growers—who were not growing over 50 bushels of corn an acre by applying from 500 to 1,000 pounds of dry fertilizer per acre—apply from 8 to 12 tons of limestone per acre and increase their corn yields from 65 to 175 bushels. Of course, this could have been due to many things, but in each case the application of limestone was directly or indirectly responsible for improved yields.

Muck soils are the most interesting to work with. They have a high percentage of organic matter. The surface, from 4 to 6 inches, may be very active chemically; whereas the material from 6 inches to 6 feet deep may still be in a pickled condition. If muck beds are surrounded by soils of limestone origin, they probably are quite fertile and are fairly well saturated with calcium. If, however, they are surrounded by naturally acid upland soils, they are of little value for crop land until they have had heavy applications of pulverized limestone.

A greenhouse rose grower suggested a peat problem which is interesting in terms of base exchange capacity. He had a muck soil taken from an old mill pond left lying for fifty years or more. There were 8 to 10 feet of this muck. It was mealy and loose, with no identifiable fiber, and seemed a perfect supplement to make up a greenhouse bench soil. When he mixed it with some mineral soil—1 part muck, 2 parts cow manure, and 3 parts field soil (silt loam)—the rose plants would not grow in it. I tested it for base exchange data and couldn't believe that a muck should need so much limestone. I initiated an experiment with it in 8-inch clay pots. The results are shown in Table 3. I planted lima beans in each treatment because they are very sensitive to calcium deficiency. According to my test, a 7-inch layer of this muck would require 40 tons of limestone per acre. I planted five good seeds in each pot.

TABLE 3

GROWTH OF LIMA BEANS IN MUCK SOIL

Treatment	*Percentage Germination*	*Subsequent Growth in Inches in Four Weeks*
Nothing added	0	0
10 tons limestone per acre	25	1 to 2—abnormal
20 tons per acre	66	1 to 3—fair
30 tons per acre*	50	3 to 4—fair to good
40 tons per acre	100	6—normal growth

* Apparently some injured seed.

I told the grower to add 7 to 8 tons per acre of limestone per inch of muck. This was 50 tons per acre. He measured off an area in his field which would give him sufficient soil to fill his benches. He covered this with 2 inches of muck and 3 inches of cow manure, then added the limestone and worked it into the soil with a rototiller. When he was through, he had between 8 and 10 inches of loose soil. After it was placed in the benches, I checked the number of empty bags from the limestone he had applied. It came to over 200 tons per acre. He had miscalculated and then had some misgivings as to whether he had used too much. But he set rose plants in the beds; and I have never seen roses grow so fast. People came from all over the state to see these roses. After that experience, I recommended heavy applications of limestone on all muck soil where the test showed a need for calcium.

Muck soils are usually poorly aerated because the water table is too high. This not only brings about poor aeration but also prevents the temperature from increasing. The result is a cold subsoil which tends to keep the roots from penetrating deep enough to take advantage of the nutrients. Instead of having good oxidation, we found much fermentation, which produced gases that were toxic to the roots. In this case, a hole was dug 2 feet deep and a piece of lighted paper was dropped into it. Immediately there was flare-up from methane gas, which is a product of fermentation. Later, I found there was a relationship between the formation of methane and calcium saturation. If the active organic matter was well saturated with calcium, there apparently was a change from fermentation to oxidation, because methane gas was not formed.

I had occasion to investigate a muck soil which supposedly had been over-limed. Two tons of hydrated lime had been applied on a strip of celery land. After the celery was set in and started to grow, it developed a yellow color and became stunted. According to my calcium test, there was insufficient calcium. The root growth was not normal. I applied 10 tons of limestone on a strip and worked it into the soil. It corrected the deficiency and the celery grew much better than where limestone had not been applied. The roots grew deeper. I believe the 2 tons of hydrated lime were too active in the shallow layer of surface soil, releasing too much calcium, which was not in equilibrium with the base exchange complex. It prevented the roots from absorbing necessary nutrients needed. The roots were stunted. When the limestone was applied, it conditioned a large rooting area with a saturated calcium colloid. It is important that a pulverized limestone be used for this purpose. The finer the particle, the quicker the calcium becomes part of the base exchange complex.

Each of the above soil types has modifications. In the coastal plain, there is much fine sandy loam which is not well aerated because of the fineness of the sand. It requires more limestone than the regular sandy loam. There also are poorly drained soils. They have more organic matter, because oxidation is slow, and, even with higher temperatures, a certain amount of fermentation takes place. Subsoiling these soils can improve drainage and aeration and greatly improve yields. These soils are often farmed by the ridge and furrow method, which helps drainage and aeration.

The general productive capacity of a given soil depends, at least partially, on its location, because each location may be affected by different geologic and erosional influences as well as by the prevailing rainfall and temperature. Heavy rainfall causes surface erosion, the severity of which is related to the degree of leaching. The amounts of fertilizer needed vary accordingly. Fertilizing coastal plain soils is a different matter from fertilizing soils in southern Wisconsin or Kansas. Coastal plain soils are subjected to heavy rainfall and, therefore, we expect more of the available plant food to be lost. However, even these heavily leached soils have the capacity to produce high yields with amazingly small amounts of applied fertilizer when the base exchange capacity is properly saturated with calcium.

As we go west from the East Coast, we expect more native fertility, because the average rainfall is reduced. As we get closer to our western mountains, we eventually arrive at the point where salts accumulate and, even though we have an abundance of fertility, crops won't grow because water is the controlling factor. I tested a sample of soil in the Red River Valley. I told the farmer to bag it and sell it for commercial fertilizer. He said he had seventeen feet of it.

Crop production the country over is not correlated with the fertility. Rather, yields in most states can be correlated with the activity of brains of people responsible for research work in the respective states. Most of our theories on soil fertility are not too well established and fertilizer recommendations are certainly not made from a knowledge of soil chemistry. Most recommendations are not based on much more than farmers' practices or on some testing experiments. In most states we have wide variations in levels of soil fertility but only one set of recommendations, which in many cases are suggested by fertilizer salesmen.

Any soil is made up approximately of the same ingredients, but they vary in proportions. In Table 4, I have given an approximate composition of the soils that are farmed. Sandy soils

TABLE 4

DIFFERENT SOIL TYPES

	Fine Clay	*Clay*	*Silt*	*Fine Sand*	*Sand*	*Organic Matter*	*Buffer*
Sand	T	T	T	L	VH	VL	VP
Loamy Sand	T	L	L	L	VH	L	P
Sandy Loam Light Phase	T	L	L	L	VH	L	L
Sandy Loam Heavy Phase	L	L to M	M	H	L to M	L to M	L to M
Silty Loam	L	L to M	M	L to M	M	L to M	M
Silty Clay Loam	L	M	H	L	L	M	M
Clay Soils	M	H	M	M	L	L to M	H
Clay Loam	H	H	L	L	VL	M	VH
Prairie Soil	M	M	H	L	L	H	H
Muck	L	VL	L	L	L	VH	H

H–high M–medium T–trace
L–low P–poor V–very

all across the country vary in their proportions of fine to coarse sand. Coastal plain soils have more fine sand and more silt and clay in the subsoil than the sandy loams in the Lakes regions or the Central States.

Aeration, drainage, and cultural practices vary from area to area. In the East many of the sandy soils have clay and silt in the subsoil. In the Central States, many of the sandy loams are underlaid with gravel mixed with silt and clay. This is very important from the standpoint of the formation of plow soles and hardpans.

Where rainfall is comparatively heavy, rather extensive and troublesome plow soles, not hardpans, are formed. These are of geologic origin, formed long before the soils were farmed. The sand remains in the surface soil while clay and silt, because of their unsaturated condition, move toward the lower levels, where they form into layers 2 to 4 inches thick. They are very common in the coastal plain soils and adversely affect yields. Farming practices have accentuated the conditions. Plowing year after year at the same depth tends to do the same thing that continuous troweling does to the surface of a concrete walk.

The fertility levels of the sandy soils vary markedly in different localities. Their chemical condition determines their fertility level. I have had many arguments with people who are supposed to know soil chemistry. They claim that sandy soils are our poorest soils because the crops tend to have a yellowish appearance and will exhibit more deficiency symptoms. I am concerned about volume of food productivity, and it has been my contention that sandy loams are our most productive soils because the buffer system is low. Therefore, we do not have to know as much soil chemistry as we do when we deal with the highly buffered soils which contain much more chemically active clay and organic matter.

Sandy loams are better-drained, better-aerated, have less organic matter (an advantage until we know more about organic matter), and warm up quicker in the spring, which starts microflora activity. Their wilting coefficient is low, an advantage when you must depend on low, spasmodic rainfall. I have grown 200 bushels of corn on sandy loam soils, which I have never been able to do on heavier, supposedly more fertile soils. I feel that I know more about the chemistry of sandy soils because there is less to know about them. They are closer to sand culture, a simplified system of growing plants. I have spent much time carrying on sand culture experiments and have grown crops as good as I ever have on soils—because the process is far simpler since there is only the plant to consider. The feeding of the plant can be controlled according to the available sunlight. This can't be done at the present time with crops grown on heavy soils. The more clay and organic matter a soil has, the more complicated it becomes. A greenhouse grower who used the sand culture method once told me that he would use the method until he learned enough about his soil to grow as good or better roses with some other method.

Hilltops in West Virginia can produce crops as good as those of the prairies of Illinois. If the slope of the hills is such that one can farm them with tractor-drawn equipment, one can, by subsoiling the tops and applying limestone when needed, produce exceptional yields. When one uses a subsoiler and tears up by circling the top of the hill and going crosswise to the slopes, one can cause the rainwater to soak into the subsoil, where it can be stored for the future use of the growing crop, instead of having it run down the slope and carry soil and nutrients away. The limestone should be applied first, so it is partially distributed from the surface down to the 2-foot depth. The limestone helps to make the soil more mellow. Then the water will soak in more readily. I have seen this practice make it possible to produce over 100 bushels of corn on the top of hills and on slopes that were almost too steep to farm.

If you farm black prairie soil, which potentially is 300-bushel corn land but is only producing 65 bushels, you have a problem which becomes very embarrassing. Here there are tremendous quantities of plant food materials, and yet the crop is not producing as much as the submarginal hilltops of West Virginia or eastern Ohio could, if properly treated. The low yield can't be due to a lack of fertilizer. It must be due to some interference, deficiency, or excess of the wrong chemical.

The main problem that we have on the black prairie soils is a set of factors favorable to the production of leaves and stems—in other words, we have an overabundance of nitrogen. Yes, we have plenty of fertility, but the balance is all in favor of succulence of growth; an overpowering effect of nitrogen against the storage of starch and sugar, which we need for yields of grain. Grains contain a high percentage of starch. Thirty-five pounds of the dry weight of a bushel of corn is starch.

Muck soils offer still greater problems. I have seen some very high tonnages of corn silage, hemp, potatoes, cabbage, alfalfa, and many vegetable crops come off an acre, but by and large our yields from muck soils leave much to be desired. First of all, there are many different muck soils. I worked with two of these muck soils in Wisconsin. One muck had its origin in cattails, the other in grass. The cattail muck was deficient in potassium (check plot 12 inches tall; potash plot 10 feet tall) while the other was deficient in phosphoric acid (check plot 18 inches tall; phosphate plot 8 feet tall). The degree of decomposition of peat to muck of course determines its productivity. The degree of change from peat to muck or humus determines the amount of base exchange it will contain, which in turn determines the lime requirement.

A general idea of soil composition should give us a good idea of what we have to contend with. (I discuss this at this time because nitrogen has a tremendous influence on the availability of the soil ingredients which are needed by the plants we grow.) A soil is made up as follows:

STONES Round field stones of different sizes. Red shale in coastal plain soils, which usually need lime. Limestone and sandstone shale, a source of some plant food material.

GRAVEL Varies with different locations. In glaciated areas we have gravelly loams and gravelly silt soils. The gravel is a mixture of different rocks which represent many different minerals and are the source of much plant food materials, such as potassium, iron, sulphur, calcium, magnesium, phosphorus, manganese, boron, and many others. In unglaciated areas we have very little gravel, although we may have the minerals present in much smaller particles.

SANDS *Coarse sands* are prominent in glaciated areas and are washed out and carried by water. *Fine sands* are the result of further weathering and other agents of erosion. In coastal plain soils, fine sands may become poorly aerated and may require the same treatment that silt and clay soils need.

SILT Finer than sand but coarser than clay. Found in all soils. A source of many minerals used by plants.

CLAY Finer than silt. If fine enough becomes a chemically active colloid which does not dissolve in water. The basis for base exchange and buffer activities in the soil. This makes the difference between sand culture and soil culture. All our soils except the purest sands have some. I like to think of it as a jelly-like film surrounding the larger soil particles. The physical condition of this film, or colloid, as most scientists call it, has much to do with the magnitude of the yields that may be expected and probably is as much to blame for our abnormal soil problems as anything, except organic films, which may also become colloidal.

ORGANIC MATTER The least understood of the soil ingredients. If undecomposed it is not chemically active and probably serves no other purpose than to loosen the soil and permit better aeration. If once decomposed it releases nitrogen and other plant foods, humic acid, proteins which are chemically active, fats, oils, organic acids which are chemically active, starch, sugar, and fiber from which humic acid is derived. The reaction of the soil varies as organic matter is decomposed, and speeds up or slows down the process under any given set of conditions provided by temperature and moisture. Oxygen is necessary to decompose organic matter. It may be a controlling factor in plant growth in muck soils.

MOISTURE Water held in the soil as capillary, hygroscopic, or free water. This moisture makes up the soil solution and determines whether the soil is productive or not. If the soil has too much water, so that it interferes with the ready movement of air, the soil becomes waterlogged and unproductive. Roots suffocate with poor drainage.

AIR Every productive soil must have air. Oxygen is the life-blood of any soil and determines the volume of growth of a crop. It is necessary to break down organic matter and for all oxidation processes. Oxidized products in the soil are efficient growth promoters. Unoxidized products in very minute quantities may be efficient plant growth promoters but they soon become toxic as they increase in concentration in the soil solution.

GASES With good oxidation the gases in the soil are restricted to carbon dioxide. The odor of the soil is very savory and clean. Some of the carbon dioxide forms carbonates with limestone and carbonic acid, if the soil has too little base material to neutralize it. Some of the carbon dioxide comes out of the soil to join that which is given off by the plants during the night as a result of the breakdown or release of energy from sugars, starches, and other similar products. With poor oxidation or reduction numerous gases are formed. Carbon may form methane (marsh gas). Proteins break down to nitrogen and ammonia, although the ammonia usually comes from nitrates or nitric acid. We may have hydrogen sulphide—a rotten egg odor—commonly found in septic tanks. If you smell a handful of the soil, you will find it has a decayed odor.

SAND CULTURES CAN TELL US MUCH ABOUT SOILS

From sand culture (washed white sand) experiments we know that we can grow a good plant by supplying the nutrients needed by plants through water, and we can accomplish our purpose with much less plant food material than many of us think we should supply. If we use red sand instead of white, we get phosphoric acid deficiency symptoms. Red soil is more difficult to manage than yellow or white sands because we have to add much more limestone to prevent the iron in the red sand from inactivating the phosphorus ion. So far, we have had no problem omitting organic matter; we have had 100 per cent water-soluble nutrients; we have had good aeration; and the balance between the nutrients has been favorable. The acidity of the nutrient solution applied was between a pH of 4.0 and 6.8, without any harmful effects; but when the pH rose to 7.2, we began to see iron chlorosis making its appearance. The addition of large quantities of ferrous sulphate, poured over the roots, and sand seemed to alleviate the deficiency only very slowly; whereas if we put the iron sulfate in an atomizer and sprayed the foliage lightly, it corrected the iron deficiency overnight.

From this we can assume that at the higher pH the iron was inactivated in the sand culture and could not be absorbed by the plant. But what about the foliage spray? This opened an entirely new field of applying fertilizer, which we will discuss in a later chapter.

So we go back to our sandy loam soils and we grow crops with indifferent results, until we check the calcium. We find that as long as our calcium is not high enough we can't get good yields, even though we supply large amounts of commercial fertilizer. In other words, the base exchange complex, which we know too little about and which is a godsend in our soils, is acting as a stumbling block to the nutrition of our plants. If we could remove that and organic matter, we would be back to our starting point—sand culture. Thus, what we have in our soils is a hindrance. That remains to be seen as we find out more and more about the different soils. We do know that simple sandy soil will grow plants very well providing we supply moisture and nutrients. As we get into our complex soils, we have to learn how to deal with extraneous matter rather than with what we have to add to the nutrient supply. Since soils have a complexity of materials which we can't remove, our problem is to inactivate their interference, and limestone seems to do this more effectively than anything else.

I built a plant grower—a wall-out of limestone rock 3 feet tall around two sides of my porch. When we had it finished, my mason said: "I have several loads of subsoil that was left when we dug our basement hole. It isn't much good but it might do to fill in two feet." So, we got a load. From its appearance, any sane person should have used it for road fill. It was silt and clay, crumbly, yellow, red- and blue-mottled in color, with pieces as hard as brick. I looked at him, then at the soil, and said: "Okay, we will use it, but I want to mix plenty of limestone with it." We filled in 4 to 6 inches with this subsoil and put an 80-pound bag of limestone on it. This was an area 2 feet by 24 feet, or 48 square feet. When we finished, we figured we had not less than 50 tons per acre of limestone on each 8-inch level of soil, and it was well distributed. We finished the load and had another 12 inches to fill. He said: "What are you going to use for the top?" I said: "Another load of the same stuff."

We finished the filling on Saturday evening. On Sunday morning, when the soil was still sticky from the water I poured on the previous evening, the second of August, my wife gave me odds and ends of plants from the garden, which I set in. It seemed too late in the year to expect anything from them; but by the time frost killed most of the plants, they were a sight to behold. It was a conglomeration of plants 2 to 3 feet tall and full of flowers. No fertilizer had been applied. I did expect that the aeration the soil got from being exposed to the air, with the limestone, would grow good plants, but I did not expect much luxuriant growth the first year.

On the basis of these observations, it would seem that our problem is to reduce interfering influences rather than to add something that will make the plant grow more efficiently. Every soil has chemically active clay and organic matter which in an unsaturated condition exerts more osmotic pressure than the roots of plants can exert, with the result that soluble nutrients may be drawn and held by the base exchange complex to the disadvantage of the growing plant. This hypothesis is not valid unless we assume that some oxygen plays a part in the interchange. Certainly oxygen plays a very important part in the type of growth that is necessary to produce a big yield of corn.

According to research work done by Europeans early in the twentieth century, the calcium ion plays an all-important role in the yield potential of any soil. This apparently is accomplished by having sufficient calcium supplied to the soil, so that 80 to 85 per cent of the negative ions in the base exchange complex are neutralized by the calcium. This neutralization process apparently relieves the osmotic pressure so that the plant can abstract the nutrients it needs. We must remember that an application of barnyard manure can build up additional base exchange, and it is reasonable to assume that the 25 additional bushels of corn obtained when 300 pounds of limestone are added to a yard of manure may be due to the neutralization of active organic matter introduced with it. Even plowing under a heavy green manure crop may temporarily increase the calcium requirement of a soil.

Most research thinking has been done in terms of supplying more plant nutrients or introducing factors which eventually will add additional nutrients or build them up by resting the soil. The calcium needs of a crop are not large, but the soil needs are high and vary with the soil type. As long as the calcium is maintained at a satisfactory level the soil will produce good crops.

One hears the expression "worn-out soil." I don't know what is meant. I have an idea it means "depleted of nutrients." It is an expression commonly used by fertilizer people to sell fertilizer. I have seen soil made very unproductive by the continuous use of commercial fertilizer. The grower, upon good advice, made the soil productive by adding limestone and withholding fertilizer for several years. That seems to belie the idea that soils wear out if you don't apply quantities of commercial fertilizer. The above should not be considered as evidence that fertilizers are no good. To me it means that the people who advised this farmer on his fertilization program didn't know what they were doing. I shudder to think of the information farmers are getting from people who are supposed to know.

In conclusion, I would like to note the pointlessness of some of our farm programs. As a result of fifty years of testing research that has barely maintained our yields, we have put the cart before the horse. I am giving some farmers credit for learning how to grow good crops by trying new things. With the soil situation we have in the United States and practices based on the classification of good, fair, poor and submarginal land staring us in the face, it amazes me that we are still maintaining surpluses of some crops. It is my honest experience that it is easier to grow big yields on poor and submarginal land than on so-called good soils. I have worked with farmers in many states who will support my thesis, because of the results it has obtained.

CHAPTER 7

Our Commercial Fertilizer Research Program Is Not Tenable

FERTILIZER COMPANIES were formed on the supposition that every farmer needs fertilizer to grow crops. This was based on the idea that you could deplete the fertilizer (plant food) in the soil quickly by continuous cropping. From the dealer's point of view, when you form a company to sell fertilizer, you expect not only a comfortable salary but profits for the stockholders. A little arithmetic showed that the potential output of a factory could be very lucrative even if each farmer used only 200 pounds of fertilizer per acre, and many mixing plants were established to assure farmers within a short radius fertilizer when they needed it.

Research work was running a bad second to fertilizer salesmen, who were inducing farmers to buy commercial fertilizer before much proof was available that the fertilizer was needed. After some fifty years of usage in some areas, research work has lost ground in the race between volume sales and established research facts. The value of fertilizer for producing more crops per acre is still very much confused, because the problem is very complex.

The early work considered replacement of barnyard manure, or supplementing it with commercial fertilizer, because there were visions of horses being replaced by tractors, and a more clear-cut division between animal farms and grain farms was in the offing. This has been carried to the point, in some cases, where we have so many animals per acre that farmers have too much manure for the good of their crop yields and grain farms.

Along with the use of commercial fertilizer, there was much discussion about the use of a single ingredient. Superphosphate and rock phosphate came in for much discussion as a supplement to manure and, because superphosphate was more soluble, it was considered a better product. We still argue about this. People who are controlled by the fertilizer industry naturally are interested in furthering the sale of the soluble product—superphosphate. Imagine my surprise upon finding a bulletin put out before 1920 by an eastern experiment station to the effect that superphosphate was toxic to plants. I doubt whether the bulletin was ever distributed. The more I studied the bulletin (based on work done by an authority who was a friend of mine) and talked with the author (it had become a distasteful subject to him), the more I became critical and careful about the use of superphosphate. There probably has been more research on superphosphate (acid phosphate) than on any other fertilizer ingredient—and we probably know less about it. My humble opinion makes me wonder why we know so little about its relation to nitrogen and potash. From my experience the interrelationship between nitrogen, phosphorus, and calcium is extremely important in our food production program. My experience is that unless you know what you are doing, there are conditions under which superphosphate may be toxic to crop plants. Since my start in fertilizer research, many fertilizer ingredients have come on the market which have tended to increase the solubility of phosphates, and I think they have confused us rather than helped us. There is no comparison possible between ammonium phosphate and superphosphate, unless you know how soluble they are in the soil when they are applied and what effect the gypsum in the superphosphate has in promoting yields. Many of our early experiments can be thrown in the wastebasket, because there were too many variables which were not checked.

If we check the experiments carefully, we can't help but wonder how much of our research has contributed to the knowledge about what part commercial fertilizer actually plays in our food production programs. Frankly, I don't believe that fundamentally we can give fertilizers too much credit, in spite of the billions of tons used by American farmers. This is not a criticism of their place in our food production program, but it is a criticism of the type of research that has been acceptable during the past fifty years.

If one follows the comments in farm journals, items in the daily press, and discussions in national committees and farm discussion groups, it seems that they all work on the assumption that the basis for all crop production is fertilizer, with an occasional side remark that limestone may be helpful. This is a ridiculous assumption which is based largely on propaganda and hearsay. What amazes me is that the majority of our agronomists sit by without a note of criticism.

The first year that I was established in experiment station research work, Dr. Wheeler, who was then Director of the Rhode Island Experimental Station, told me that it was all right to work on fertilizer but that I shouldn't underestimate the value of limestone in the soil. I have realized more and more that he gave me more to work on than any man I came in contact with and, if I were to make one criticism of our early fertilizer experiments, it is that the importance of the calcium ion was overlooked. This is puzzling because of the very enlightening work done by Ganz, Way, Hissink, Gedroiz, Kelley, Jenny, and others before 1930. As far as I am concerned this omission relegates those experiments, including some of mine, to the wastebasket. In my humble opinion no fertilizer experiment should be initiated until the base exchange complex in the soil is first properly saturated with calcium or is part of the experiment. In my early experiments, I made the mistake of depending on the pH test to tell me the calcium saturation in the soil. Today my potentiometer is dusty from many years of idleness. Since I have depended on the calcium tests, my experiments have shown results that seem to correlate with soil conditions. Furthermore, I get 40- to 100-bushel yield differences with fertilizers where formerly I had to analyze my data statistically to find out whether my 6-bushel increase was a significant difference. I am of the opinion that when you have to analyze data statistically to find out whether you have worthwhile difference, you may as well throw it in the wastebasket. We have a lot of statistically analyzed data published in our journals which has contributed nothing to our knowledge of how to keep our population from starving.

There is nothing wrong with statistical methods. They are based on mathematics, a fundamental science. My criticism is that too many of us use them to prove the value of data that was collected from an experiment which was initiated on the wrong premise. Too many of our fertilizer experiments were started on a faulty premise.

I have argued with colleagues on many subjects and I think they thought I had radical ideas. One of them told me, "I expect you will argue with the Grim Reaper on your deathbed." In my early years of research, I wasn't quite sure; but as I gained experience and found that I could take corn land producing 50 bushels of corn per acre and increase the yield to 150 bushels with my idea of using fertilizer, I became convinced that my radical ideas were on solid ground.

Because of my unorthodox methods I have few county agents, agronomists, and Vo. Ag. teachers listening to me. They say I am wrong, and yet they can't increase yields above 60 bushels on plots where I get 100 bushels or more. I do have a large number of farmers who have taken my ideas and accomplished the same things I have, and when they invite a county agent in to see the results, his only comment is that it won't work. He doesn't trust his eyes, let alone his thinking.

There are some open-minded county agents and high school teachers who work with me. It makes a difference where they were educated. For some reason those who come from western or foreign universities are more tractable and far more open-minded and seem to be able to do their own thinking.

When I taught juniors and seniors in college, I tried to teach them to think for themselves—for several reasons. Nothing that I was teaching was so well established that students should spend their time memorizing it, and secondly, the biological field is so variable that every problem a student faces when he is in the field is relative to something else. I was surprised that so many students wanted to memorize everything. They hadn't had enough training to work things out for themselves.

My boss used to give freshmen a talk when they entered college. His theme was: "How adaptable are you? In a changing world you have to make decisions every day. How well educated are you to make those decisions?"

I give lots of talks to farm groups. I try to give them an educational talk to help them better understand their production problems. In recent years I hear county agents and fertilizer salesmen who are primed, are anxious to get me in a corner by asking me embarrassing questions. I usually direct them by invitation into the front rows so they will have ample opportunity to ask their questions. They usually sit and listen from 10 to 12:30 A.M., but never ask any questions. One county agent, as he left the room, was heard to say, "I still don't believe him." One fertilizer salesman came to me after the meeting and apologized for having said publicly said "Tiedjens doesn't know what he is talking about." Then he added, "I have never heard a fertilizer talk that was so interesting and made so much sense. From now on, if I hear of your meetings, I hope I can attend them." As I said before, my ideas are different from what I was taught in college. They contradict presently held ideas, but every statement I make is based on what I have observed while working with farmers in twenty-three states.

Not all farmers agree with me, and some won't listen to me. They talk with their county agricultural agent and because he is a government official, they assume his word is gospel even when they can't grow over 50 bushels of corn with the advice they get from officialdom. Instead of suspecting their methods, farmers are led to believe that they are unfortunate in that they are located on submarginal land.

After I spoke to a farm group in southeastern Ohio, a man who looked like a bank executive came up to me. He said, "I am one of those unbelievers. I felt that these submarginal hills, as they are called by the experiment station people, were good for something, and I bought quite a few acres. They are rather steeply rolling, and I started applying limestone and subsoiling around the hills, as you suggested today. But I did it for a different reason. I wanted to stop erosion. I did not think about storing that water in the hills. I put lime on to grow clover. I had always heard that you didn't need limestone for corn." I told him he didn't if he was satisfied with 50 bushels of corn to the acre. Then he told me he tried growing corn on one hill and harvested over 100 bushels. Since that, he has grown corn on several, and he always gets over 100 bushels. He said, "My experiment station friends come out here, look at my crops, and shake their heads and can't understand how I do it. After listening to your ideas, it is quite obvious why I get those yields. So far, I haven't used any fertilizer. I believe I will start to use some. Perhaps I can double my yield."

Five years later I visited his farm by accident. My salesman wanted to show me a good field of corn. I heard later it averaged 137 bushels. It was a good yield for submarginal land and since I saw the field, I know I was misinformed about it. I also heard that the corn in the black ground in the valley nearby only made 71 bushels but not because of any deficiency. From the appearance of the foliage, I assumed it had been oversupplied with nitrogen.

Agronomists have been my greatest hecklers. They seldom attend my meetings. Those that do are usually friends of mine who are open-minded enough to want to learn. I have talked with agronomists who give you the idea that all knowledge comes from them. I know farmers who have more common sense than some agronomists with doctor's degrees.

If a person feels that he is through learning when he has a doctor's degree conferred on him, he has lost his usefulness to society. If his education has done him any good, he should be humble and be more eager than ever to seek the truth. I sat in a lecture at Harvard University on personality and reward, and I heard the following statement made: "The unfortunate result of granting doctor's degrees is that so many graduate students make a worthwhile contribution to science in their undergraduate years, but after they have been granted a degree they seem to forget their obligation; you never see any further contributions bearing their name."

I have thought about this a great deal and I wonder whether there may not be an explanation other than that inferred by the Harvard doctor. I had a man in my employ who had finished all his undergraduate work for a degree from the University of Minnesota but had not finished his thesis when I hired him. He was very humble in his thinking. Some people said he was lazy, but I did not agree. He had what I thought was a really new approach to his field of major study. It was quite different from his material for his thesis. He had lost interest in his thesis because he became so enthusiastic over his new approach. I spent a good many hours discussing this "baby" with him. Because it was a new approach it overwhelmed his thinking by opening up so many avenues of approach. I practically stood over him for two years before he finished his thesis so that he could qualify for his degree. I am doubtful whether he will ever publish anything along his new way of thinking, because he has gone into extension work and probably won't have time to work on the ideas that he presented to me. It takes a lot of people to make a world. Fortunately, they are not all alike. If they were, we could not argue. Constructive arguing is educational. But when it comes to fertilizer research, it is easy to get our thinking into a rut. I soon found out that there were a lot of loopholes in our thinking which in many cases could be used to someone's advantage in the sales field.

I approached my first research project with much enthusiasm. This was at the beginning of the 20's. I made the acquaintance of many fertilizer company representatives, some of whom had very definite ideas on fertilizers, while others were simply holding a job.

I shall always remember the advice one experiment station director gave me: "You will find many people in the industry. Don't let them influence you unless they know what they are selling." After I got into my own company, I decided that the worthwhile salesman was the man who knew what he was selling.

When I first went with the New Jersey Experiment Station,

I was called to a farm in eastern Pennsylvania by a fertilizer salesman. A 10-acre field of horse-radish was not responding to fertilizer. I was told it had a creeping disease which started in one corner of the field and was gradually sterilizing the field. Plant pathologists had been working on the problem for three years and had found no disease organism. In one row the 4-inch root cuttings hadn't grown a single leaf, while in the next row the same size of cuttings had grown into plants 2 to 3 feet tall.

The reason they asked me to look at the field was because the grower questioned whether fertilizer could be doing the damage, since the sterile soil spots were appearing all over the field. I asked about the lime on the soil and found the pH was 6.8. This meant nothing to me because I never depended on this test to determine whether lime was needed. I took some soil samples and found that available calcium in the sterile soil was non-existent, while in the good areas there were 100 p.p.m. Calcium was on the threshold of being deficient. We corrected the problem by applying 4 tons of pulverized limestone per acre. The "disease" completely disappeared.

After I saw the results from the limestone on this soil, I thought of my asparagus experiment. I had tested the seashore sand on Cape Cod and found it to be near neutral, so I had assumed it had sufficient limestone. I am sure now that my problem was calcium deficiency. Plowing the carrots and grass under preserved some available calcium, whereas the rutabagas, which need an appreciable amount of calcium, took the calcium to market with them. Since then, I have had many similar cases where a heavy application of limestone immediately and materially boosted the yield of potatoes, corn, sugar beets, soybeans, and tomatoes without adding any additional fertilizer.

A friend of mine, who farms considerable land in Fayette County, Ohio, told me he didn't feel that he was getting much good out of the commercial fertilizer he was using and he wondered whether the stuff was any good. I told him the state inspection service saw to it that a labeled 5–10–5 fertilizer couldn't be anything but a 5–10–5 and that the trouble probably was in his soil. He could imagine that, since he applied several tons of limestone in certain other fields with good results. Whenever his trucks were not hauling for the neighbors, he was applying limestone on his own fields. It was the old story of the shoemaker's children.

We finally agreed to run an extensive experiment on a 53-acre field which he was planting to corn. The rows were 800 feet long. We laid off four series of treatments crosswise to the rows and 200 feet long. We put 4 tons of limestone on the first series, 1,000 pounds of 3–12–6 fertilizer on the next series, nothing on the third series, and 2 tons of limestone on the fourth series. Then we plowed it and planted every eight rows with a different treatment, as shown in Table 5. Actually, the only thing that

TABLE 5

YIELDS OF CORN ON PLOTS TREATED WITH DIFFERENT FERTILIZERS AND LIMESTONE

	Yields in Bushels per Acre			
Treatment	*Series 1. 4 tons limestone plowed under*	*Series 2. 1000 lb. 3–12–6 plowed under*	*Series 3. Nothing plowed under*	*Series 4. 2 tons limestone plowed under*
1. Dry fertilizer in row	126	87	101	127
2. 2 gal. 10–20–10 solution in row	137	128	141	129
3. No fertilizer in row	121	119	126	117
4. 4 gal. 10–20–10 in row	130	123	138	128
5. 8 gal. 10–20–10 in row	117	112	113	112

showed any appreciable effect on the yield was the 4 tons of limestone plowed under. That the fertilizer solution was better than the dry fertilizer was probably due to the trace elements included in the mixture. When the plowed-under fertilizer actually decreased the yield he said he would probably get the good the next year.

But the experiment was run in 1953 and every year the series where the fertilizer was plowed down has produced the least corn. The reason is lack of calcium in the base exchange complex. In his other fields he has applied as much as 10 tons of limestone per acre, and his yields are in the area of 150 bushels of corn per acre.

There is an understanding among agronomists that 100 bushels of corn need 200 pounds of nitrogen, 150 pounds of phosphoric acid, and 200 pounds of potash. These are not in my book, but I will accept them for the sake of argument. Agronomists intimate that if you want to grow 100 bushels of corn, you must add the equivalent amount of fertilizer to your soil or your soil will wear out. There is no argument that a corn crop needs a certain amount of plant food material, but there is no proof as to the approximate amount necessary. The federal government, in Bulletin 369 in 1941, showed there was nothing definite in the literature to prove it.

When it comes to intimating that we must apply the equivalent amount of fertilizer that the crop needs, I can't buy it. There is no known proof for this. If you apply 300 pounds of fertilizer and get an increase in yield of 20 bushels it doesn't mean that the next 300 pounds will give you an additional 20 bushels. I have conducted enough experiments to know that there isn't anything sure about it. It is largely propaganda put out by fertilizer salesmen. It would be convenient if we could figure our fertilizer needs with a slide rule, as some people have tried to do, but I doubt whether the correlation between yields and units of fertilizer is significant.

We know that if a corn crop is grown on a soil having an abundance of plant food materials, the amount in the plant will be much higher than if it is grown on a soil having a low level of fertilizer, and yet the yield of grain may be bigger on the latter field.

We apply large quantities of potassium on the soil in some areas and we do get responses, but it is not clear whether the crop benefits directly or whether the increased growth or yield is due to something else the potash may have released. I say this because of some work I did a number of years ago. These results were not published because they seemed contradictory. Now that I am more familiar with the changes brought about in a soil when you add a chemical, and because the same results were obtained on alfalfa by a midwestern university, I am convinced they are highly significant, and I am presenting them in Figure 2.

FIGURE 2

RELATIONSHIP BETWEEN POTASH, LIMESTONE AND YIELD OF PEPPERS (MANGOES)

500 lb. of potash applied per acre in Series A; no additional potash applied in Series B

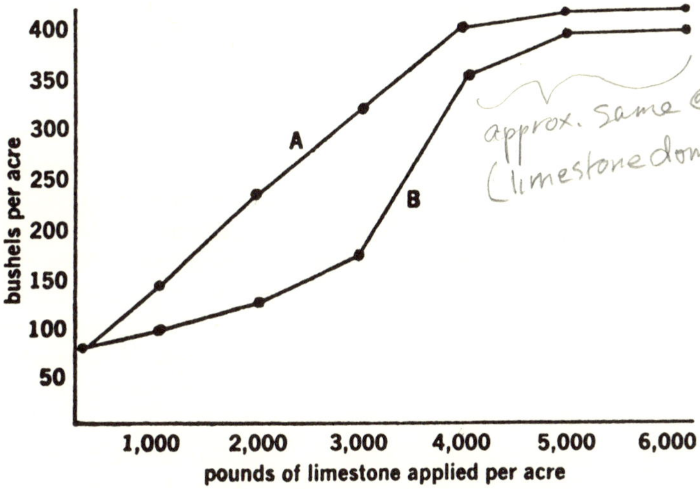

As limestone was applied beyond 2 tons, the difference in yield between potash and no potash decreased. Of course, two or more interpretations are possible. Either the limestone released potash in the soil, or the potassium replaced calcium on the exchange complex and made calcium available to the plants. According to the calcium test on this soil, 3,600 pounds of calcium was necessary, and I found only 1,600 pounds. This indicated that 5 tons of limestone was needed. By applying 3 tons, we harvested 400 hampers of peppers. This was a good yield, but I have seen 800 to 1,000 hampers harvested. Most pepper growers use too much nitrogen, which results in a big leafy plant and interferes with the set of fruit, thus reducing the yield.

It is possible that if I had used more limestone, the top yield in the experiment might have been higher. The question in my mind now is whether there would have been more difference in yield on the low limestone plots if I had used more than 500 pounds of potash.

When potash is applied to prevent deficiency in the plant, 200 pounds is usually sufficient unless it becomes unavailable to the plant because of some chemical condition in the soil. If you are applying potash to release calcium from the base exchange complex in the soil, you must use much more. Sweet potato and melon growers on coastal plain soils applied as much as 1,500 pounds of muriate of potash on an acre and claimed they got a good response. I doubted this and repeated the treatment by applying some limestone and no potash and found the high potash application reduced sweet potato yields by from 50 to 100 bushels and melons by 50 hampers.

We checked the amount of calcium in the water coming out of the tile drains where the heavy applications of muriate had been made, and where heavy applications of commercial fertilizer were used we found 40 parts per million of calcium in the water. With a ton of limestone carrying 400 to 600 pounds of calcium, it is easy to figure why calcium is becoming a critical element.

Many of our experiments leave more questions unanswered than they answer. When you enumerate all the things that can affect yields of crops it is easy to see why we know so little about fertilizers and their effect on crop yields. As a result, we have difficulty evaluating the future part that fertilizers will play in crop yields and our world food supply.

Although I did much research on fertilizer, I published very few results because I could not satisfy myself that the results meant much in terms of profits to the farmer. If it cost more for the fertilizer than the increase in yields was worth, I saw no reason to give them publicity. All the propaganda being put out by the industry showing the tremendous importance of fertilizers in our future food supply made me feel that unless I could get outstanding results, the published data probably would not be worth the space it would occupy in our publications.

I did considerable work trying to find out the part that limestone plays in the use of fertilizer. Also, the placing of the fertilizer with reference to seed or seedlings took much of my time. These results were very confusing. By placing the plant food farther away from the plants I found out I could use less fertilizer and, when I plowed it under after broadcasting, I got still higher yields. It did not make sense until I found out that no fertilizer gave me my best yields—but even then the top yields were not high enough. In other words, I wasn't studying the effect of the fertilizer from the standpoint of the nutrient needs of the plant. I was studying means of decreasing toxicity; the less the toxicity the bigger my yield was, but it never got beyond the yield I got with no fertilizers.

I was involved in an extensive study on the effect of fertilizer on mosaic (streak) in tomatoes. The old practice of applying fertilizer in the row had me puzzled until I found a discussion in Russian literature of the relation of phosphoric acid to mosaic. I noticed that heavy, early-season rainfall caused a high percentage of the plants to be affected with mosaic, whereas a dry spring showed very little mosaic but considerable injury because of the proximity of the fertilizer to the roots. When I checked the nutrients in the leaves and stems, I found the phosphorus very high in the mosaic plants and normal in non-mosaic plants. In wet years more of the phosphorus was in solution and, therefore, the plants gorged on it. In dry weather very little of the phosphorus in the soil was available, and only small amounts were found in the plants.

I have had many arguments with agronomists about the availability of fertilizer when applied to the soil. Some claimed it was 100 per cent available. According to state chemists it is soluble, but what happens in a laboratory test tube and what happens in the field are two entirely different things. When I got into fertilizer solution work I found that it took a lot of stirring over a long period to dissolve a 5–10–5 fertilizer, and the more they pelleted the fertilizer the more insoluble it became.

As I got more information from papers published in Russian journals, I found that too much or too little phosphorus could cause mosaic. So, I took the fertilizer out of the row, applied it broadcast, plowed it under, and eliminated mosaic. There is more detail on this in the chapter on tomatoes and fertilizer. I never published these results because the director decided the fertilizer people might not like it. It amazed me to think that the fertilizer industry would stand in its own light. In recent discussions I have heard that they apparently now realize their mistakes and are looking for an easy way out of their dilemma.

The nitrogen industry is fast becoming a white elephant. Nitrogenous ammonia and urea and their oxidation products are easily made from the gaseous nitrogen in the air. Many chemical companies are making nitrogen compounds and are looking for sales outlets, and every person who sells them will continue to push them even if the customer has them running out of his ears. They pay no attention to the customer's needs. They will sell to the farmer who has an abundance of nitrogen in his soil as quickly as to the man who has a crying scarcity.

One of the big problems we encounter in trying to increase yields is the abundance of nitrogen in many of our potentially productive soils which are not producing at the present time. In spite of this, thousands of tons of nitrogen are being sold to farmers every year. This will further reduce yields on a high percentage of the soils high in organic matter. (See Chapter 11 on sunshine and nitrogen.)

I have had farmers who have high organic matter soil tell

me they are sure that nitrogen increased their yields. When I questioned this, they said they could see it in the greener color and faster growth. When I asked them how much increase in yield they got, they said they did not check their yield. When I asked them how much shrinkage they had in their corn cribs, they said that it shrank down about 20 inches. When I measure yields, I want comparisons made on a dry matter basis. A bigger ear at harvest doesn't mean more shelled corn. The only sure comparison is a No. 2 shelled corn comparison.

Farmers have also told me that they knew that they were getting good results from fertilizer, because if they didn't turn the fertilizer on at the beginning, they could see the difference in growth. I was standing with one by a corn field that was ready to pick. When I asked him to point out two such rows, he said, "Sure, there are two right down here," and we walked until we came to two rows in which the stalks were shorter than the others. I looked at them and pointed out that they were dead furrow rows. "Well, they must be farther down." So we walked some more. We finally came to two rows where the stalks were all of a foot shorter. "These must be the rows." We looked at them and I pointed out the fact that the ears averaged bigger than on the rows alongside. He said, "That's funny. These can't be the rows." I said, "I think they are, but you should have put a marker here." We couldn't find any other rows. But he was so confused he said he guessed the fertilizer was all right. I said, "There is nothing wrong with it, but I am wondering whether it is making you any money. It may give you a little more silage." But then, he surprised me with the statement: "But I want good ears on my silage corn. Stalks without ears don't make good silage." I told him the only way to make comparisons was to have some check plots next year, put some stakes in so he knew where they were and when he harvested them, see how many rows it took to make a load. "Well, I certainly will do that. I can't afford to use fertilizer if it doesn't do me any good."

Many of us have the idea that commercial fertilizer on our good land is the only salvation of our future food supply. Applying plant food is only a small part of our crop production problem. Commercial fertilizer, when properly used, can help increase crop yields, but if our experience during the past fifty years is an example, we haven't learned how to use it. Our average acre yields have not increased significantly in spite of the fact that farmers have spent millions of dollars for commercial fertilizer. Even the use of barnyard manure has worked very inefficiently toward increasing yields.

One might well ask, Why? There is only one answer. There are too many complications and our scientists have been satisfied to conduct their research under very limited conditions. Of course, we should not be too critical if a person approaches his research with an open mind and has the ability to co-ordinate his work with that done in other areas and other parts of the world. There has been too little integration between scientists. Too many prefer to stay in their own back yards and, I am sorry to say, I have found too many ready to look down their noses at work done in other centers of research. We should also mention that we have had too many pressure groups breathing down the necks of research men who are responsible for finding out fundamental facts.

I doubt whether many people know the real function of fertilizer. Too many have the idea that if we wish to double our yield all we need do is to apply twice the amount of fertilizer. Nothing could be further from the truth. Crop yields depend on how efficiently plants can manufacture and store sugars, starches, fats, and oils, which contain carbon. We emphasize proteins, but actually we need to have the starchlike materials before we can have proteins. In other words, proteins are made at the expense of carbon compounds in the plant.

A bushel of No. 2 shelled corn contains 15 per cent water (8.4 pounds), 8 to 12 per cent protein (5.6 pounds), ash (minerals) less than ½ pound, and carbon compounds 40 to 41 pounds. This 40 pounds represents oil and starch which the plant makes in its leaves and stems, where the green coloring matter (chlorophyl) is present. To do this the plant uses water, which it takes in through the roots and leaves, carbon dioxide, which comes from the air by the combustion or oxidation of organic matter through energy supplied by the sun. What then does fertilizer do? Nitrogen and phosphorus are used in forming proteins, while potash acts as a catalyst but does not enter into the products in the plant. How does it act as a catalyst? If the facts were known it is probably radioactive potassium that serves as the catalyst. This is a very minute percentage of the total potassium in the plant, which accounts for the fact that a plant may show potash deficiency symptoms even though there is an appreciable amount of potassium in the plant cell.

Suppose we were to consider the corn crop. Corn is a very important stable food crop. Some work has been done to develop varieties that will grow during shorter seasons. This has shoved our corn belt farther north. It not only avoids the danger of early frosts but it has made it possible to grow at lower temperatures.

Corn adapted to grow south of our corn belt is another problem. High temperatures prevent good pollination of many varieties. Seed-corn maggot, weevil, and many other insects cause much damage. We need more attention to heat treatment and storage of corn. This, along with the adaptation of varieties to our southern states, can greatly increase the boundaries of our corn belt east, south and west. We need more information on adapting varieties before we can do much about finding ways and means of increasing acre yields. This could apply to many of our crops. We have more opportunity to extend our frontiers to the south than to the north. Of course, different crops can be adapted, but to extend the limits of any given crop will require much more integrated research.

Very few people realize the importance of sunshine in our food production problem. We assume it is ever-present, that we can't do anything about it and, therefore, we need not worry about it. Needless to say, if we didn't have sunshine we would all starve. Even if we should experience a season of an exceptionally high percentage of cloudy weather, we might expect a decline not only in the quality of our food but in the total quantity produced.

So far, we have not mentioned the potentialities of our soils. If one travels cross-country by plane north to south or east to west, one cannot help but conclude that we have a tremendous acreage of land which is doing very little in supporting our national population with food, timber, or raw materials for industry.

We would also observe that much of the land which is being farmed or has been farmed is not very productive, if the farm buildings are any criterion of the standard of living of the occupants. The level land generally is being farmed, but even some of that is abandoned. We see much rolling land being farmed. There are also many more rolling hills that can be profitably farmed. The question is whether we have the know-how to do it. When most of us think of hills, we think about a lack of fertility, lack of water and accessibility. If you can't drive a tractor up a hill, you can't farm the land unless you have a mule and one-horse equipment. If you have to farm with one-horse equipment, it may be too costly. We may have to follow the example of the Chinese and Japanese and farm with terraces. The biggest expense would be to build terraces. We might have to do some government bulldozing as well as applying limestone to hills with helicopters. We won't know for certain until we try it. Many methods which at first seem impractical become commonplace with practice.

We have a lot of people living in hill country who could make a better living and maintain a higher standard of living than many now enjoy on level fertile soil. I have helped people on submarginal hills grow over 150 bushels of corn per acre. These hills were considered too poor to farm. The procedure was to apply 5 to 10 tons of limestone per acre and then, when the ground was dry, which it usually is in July and August and later, subsoil the hills crosswise to the slopes. This means pulling the subsoiler in circles or ovals around the hill, starting at the top and continuing to the bottom. Sixteen- to 20-inch deep subsoiling keeps all the water from rainfall on the high ground during the fall and the winter. It will be stored in the subsoil instead of running down the hills. Water is stored there for future crop needs. The cost runs from 21 to 30 dollars an acre, plus the expense of planting the crop. We have gotten bigger yields on these hills than on some of our very fertile, black soils. One secret about this is not to use dry fertilizers. Fertilizer solutions must be used sparingly. Twenty-five pounds an acre on the seed and 25 to 30 pounds applied on the foliage by airplane is usually enough to grow the crop.

We can increase our present food supply three times by increasing yields. Then, we can multiply that by four by farming so-called submarginal land which is now idle. And finally, by learning how to integrate all factors which affect our yields, we can increase it still further.

We have been confused by our economists because we have a surplus—apparently from using too much fertilizer, when actually fertilizer probably has had little effect in maintaining surpluses. Surpluses are local, and we should always aim toward producing a surplus. Instead of producing surpluses we are trying to legislate ourselves into starvation. Nobody in this wide world can foretell what would happen if we should have a widespread severe drought.

In production there is wealth. We don't want to produce less. We want to produce more; but we have to learn how to distribute what we produce. As crop producers, too few of us recognize quality. When we sell a crop, we want to sell every particle, whether it measures up to certain standards or not. Farmers should voluntarily sell only the good quality and feed the poor quality. Poor quality may be sold at a much lower price for feed. We can haul it back to the field to rot. Nothing annoys me more than to buy a basket of supposedly good apples and find over half of them so poor that they go into the garbage can. A grower doesn't realize how much damage he does to himself by such tactics. If the apples had been sold as seconds or culls, there would be no comment.

Even in marketing grain crops a lot of grading can be done. I have heard mill operators compliment farmers on the high quality of their shelled corn, wheat, and soybeans. That means that they handle a lot of low quality grain. It would not be difficult to require farmers to sell their crops on a quota and equality basis.

To avoid surpluses in the future, there is much that can be

done in shifting more acreage to different crops. I know farmers who were raising only enough feed for 40 steers on a four-year rotation. By gradually changing meadow and pasture to corn acreage, they are now feeding 240 head of steers, and instead of reducing the crop yields, as several agronomists with doctors' degrees predicted, their yields have increased on continuous corn ground. The fertility level in their soil has gradually increased, while their fertilizer bill has decreased.

We have lulled ourselves into a feeling of security by assuming that crop rotation was a necessity, and we have shut our minds to new facts gleaned from fundamental research. Too many of us assume we know all there is to know about a subject, and thus can't see the future possibilities. Too often, radical types of research are condemned by supposedly educated people and progress is set back fifty or more years. Our colleges and universities are to blame for this. We made more real progress before our educational systems became so well organized. Too many educators in commanding positions stop thinking when they are put into executive jobs, and everything that comes along which is not in line with their thinking is "the bunk," as one man expressed it to me. I was in educational work for many years, conducting and supervising research work, and I hang my head in shame when I think of some of the "weak sisters" who are responsible for formulating and supervising research programs. It seems as though their minds stop working when they are hired for the job. When I changed from university to commercial work, I realized how weak many of our university people are. Fortunately, we can single out some who are the exceptions to the rule and are making real contributions.

CHAPTER 8

The Farmer Is Still a Pioneer in His Profession

NOT SO MANY decades ago, tilling the soil to make a living in the United States was considered a menial job; to grow crops was considered a simple matter. If a man had a team of horses, a cow, twenty-five chickens, and a plow, he was prepared to support a family on a piece of cutover land. He cleared the land and started growing crops. He did not concern himself with fertilizers, limestone, government help, big machinery, or weed killers. His concern was to grow something to feed his own family and accumulate a little cash, which was banked in a cracked sugar bowl in the back corner of a cupboard, or under the eaves. What things he needed to buy he often obtained by bartering eggs, poultry, potatoes, vegetables, and butter at the general store located at the corner of a country crossroad. He had very little need for any appreciable amount of cash. Bluejeans, a mackinaw jacket, a fur-lined jacket and a fur-lined cap in cold regions, and mittens, heavy shoes or boots had to be bought at the general store.

Today, some one hundred or more years later, the farmer is or should become a big businessman. He has large investments in land, buildings, and machinery and needs working capital or credit. Today, he is a specialist, and he no longer thinks only about his family. His need for cash means that he must get money to pay bills. However, he still farms and sells his produce on a supply and demand market, and if he wants to live at the same standard of living his unionized cousin in the steel mills enjoys, he must make a profit from his business. This involves a tremendous amount of knowledge. His dollar has shrunk and is worth much less than his cousin's dollar. He sells on a market where he takes what his cousin is willing to pay him for his meat, eggs, vegetables, and grain crops, while his cousin has his wages set for him by the local union. Where the farmer's hourly wage depends on his ability as a manager to sell his crops for more than it costs him to grow those crops, his cousin has no worry except to carry a union card. He needs brains and education to run his business, while his cousin needs muscle, or skill in some trade, to be able to perform his work. Of course, many farmers think that muscle is a pretty important adjunct to managing a farm, unless they have the money to buy labor-saving equipment.

The federal government has decided to help the farmer because he can't make a profit from his business. For twenty or more years, we have had a farm program based on the idea that the farmer's prices should be raised to parity basis. The program in general has been a disappointment. It has returned some additional money to the farmer's pocket, but it has restricted his activities and it has fogged the real issues.

We need more production to maintain a world food supply. Even though we have been producing surpluses, which have demoralized our markets, the price of farm products is no higher than it was some forty to fifty years back. Yields of farm crops have gone up only slightly, while costs have increased by leaps and bounds.

Today our farm population is decimated because the inefficient farmer has left the farm to go into industry. Every year we have more people leaving farms, and those who remain have the job of feeding more and more people per acre of ground. Along with this our federal and state advisory organizations have appeared less and less capable of lending a helping hand. Our economists, who should be leading the farmer "out of the woods," seem to have gotten him deeper into the woods and have not been able to show him how to reduce his costs.

The future for agriculture lies in a sound research program that will show the farmer how to get maximum yields on widely different soil types at a big reduction in cost in face of possible surpluses. Many people think it is possible to reduce our acreage and thus reduce our production of crops to the point where the demand exceeds the supply. Any plan for a program to help a farmer must be based on an understanding of his problems and his thinking. The workings of farmers' minds, like everybody else's, vary between wide limits. There are no two that think alike. Each one has a different idea and, considered on a world-wide basis, it is impossible for minds to meet on common ground. But, generally, they can be judged by what they accomplish. If one were to get acquainted with one hundred farmers as he met them, and catalogue them as to how successful they were, he probably would find that ten to twenty of those were making a good living every year regardless of the kind of soil they owned or what the weather conditions were. Then there would be a group of thirty to fifty who were less successful but were living comfortably—they would make money some years and lose some in others. When you get beyond those, the people would be more suited to working for someone else. They were not making good wages for themselves. They listened to everybody and did not possess sufficient thinking power to know what was to their advantage and what was not.

You probably would find college graduates among all of them. They might all have equally good land. They might all have the same acreage. Now if you were to consider these different personalities and try to set up some government program, you would find some overly critical, some who didn't care, and some who took advantage of every opportunity to thwart the ideas of administrators to make the program work. Besides this, there are the differences between a Vermont and a Kansas personality, a hill farmer, a prairie farmer, and a farmer from southern Florida. I doubt whether there is a program that will work all over. It is my opinion that the only solution to the farming industry, from the standpoint of maintaining a world-wide food supply, is to get the politicians out of the picture and let the farmers who can do some constructive thinking work out a solution.

Personally, I feel that if a farmer has something to sell, he can, by reducing his acre costs, find a way to maintain a fairly good standard of living. If he can't grow very much, his acre costs always will be high, and no matter how much he gets for his crops he still won't make any profit.

A sensible marketing program could solve our farm problem, but not until we gradually create in our farmers a more rational way of thinking. This probably will take a better educational program than we have today. We will have to get away from selfishness and adopt a more Christian attitude toward our neighbors. Unless we can come to some agreement among ourselves to control our marketing on some sort of self-imposed quota basis, perhaps by more severe grading and feeding a certain percentage to animals, we won't be able to prevent bad slumps in market prices. Otherwise, we may turn back to peasant farming, where the farmer has little to say about his own business.

The farmer's main complaint is that it costs him more to grow the crop than he can get for it. In other words, high costs and low yields, low prices, because of surplus production, can cause him to lose his farm—regardless of support prices. Economists or other public agencies could help him regulate the movement of staple crops to market to maintain a uniform price, or the research man could show him how to increase his yields and lower his costs. In other words, he must produce corn at approximately 37 cents a bushel, so that when the price gets down to 50 cents he can make 13 cents a bushel. If his costs are 75 cents a bushel, he naturally will lose money.

The cost of growing a crop of corn—aside from the cost of fertilizer—is more or less fixed, and is more or less the same whether 35 or 135 bushels are grown. Thus, it would seem that a simple solution is to increase yields. Some farmers have done this on sound advice. If, along with increased yields, the cost per acre can be reduced, so much the better. Many factors enter into this. However, it is worth the effort. Many of my growers have realized up to $75 an acre net profit, in years when weather conditions favored a good yield.

TABLE 6

COST OF PRODUCTION ON TWO SUCCESSFUL FARMS

	Grower 1	*Grower 2*
Location of farm	Central Indiana	Near Columbus, Ohio
Soil type	Deep brownstone	Bennington silt loam
Suitability for corn	Excellent	Very poor
Yield before 1953	175 to 200 bu.	40 to 50 bu.
Crop rotation	4 years	Continuous corn
Average yield after 1953 (on 5-acre field)	196 bu.	139 bu.
Average yield on 100 acres	150 bu.	93 bu.
Highest yield on 5 acres	213 bu.	160 bu.
Highest yield on 1 acre	219 bu.	196 bu.
Number of trips over field	9	6
Cost of growing crop	$50.00	$45.00
Cost of weed control, etc.	$10.00	$ 7.50
Cost of fertilizer	$69.00	$ 9.00
Kind of fertilizer used for 7 years	Dry	Solution
Limestone applied	Some	All that test called for
Total cost of growing crop, less lime	$129.00	$ 61.50
Value of corn @ $1.00 bu.	$196.00	$139.00
Profit over cost	$ 67.00	$ 77.50
Cost per bushel	$ .66	$ .44

Examining the figures in Table 6 carefully, we find that even though Grower No. 1 produced 57 bushels more per acre, his cost to grow that extra corn was more than he got back in yield. Grower No. 2, who grew only 139 bushels per acre, had a bigger profit per acre. His corn cost him 44 cents a bushel, while it cost Grower No. 1, with a bigger yield, 66 cents a bushel. We must point out that Grower No. 1 took a big gamble on the weather. He had good weather to grow this particular crop. Had weather conditions been against him, he probably would have had half the yield at twice the cost per bushel.

Grower No. 1's yield was an average of seven out of nine years, whereas Grower No. 2 had two very wet years and very dry years averaged in to arrive at his yield. In addition to a higher cost per bushel, Grower No. 1 had to handle heavy fertilizer bags two or three times at least.

Very few farmers know what it costs them to grow a crop. All they keep track of is the money they take in when the crop is sold. If their bills total more than this, they know they lost money. This is a simple way to keep books, but it doesn't give much to work on if one is trying to figure out why he is losing money. This applies to all crops we grow.

Since the acre cost of growing corn is more or less fixed, it is interesting to speculate on the relationship between possible yields and acre costs of the above growers. Compare the costs per

TABLE 7

COST PER BUSHEL TO RAISE CORN ON TWO FARMS

Possible Yield	*Acre Costs*		*Cost per Bushel*	
in Bushels	*Grower 1*	*Grower 2*	*Grower 1*	*Grower 2*
50	$129.00	$61.50	$2.58	$1.38
100			1.29	.69
150			.86	.46
200			.64	.37

bushel in Table 7 with the selling price and figure profits. Most crops would give similar costs, except where the harvesting costs are high. In such cases, the higher the yield the higher the cost per acre would be, although the cost per basket, hamper, bushel, or ton might not be decreased by the bigger yield.

Grower No. 1 gambled on the weather, and had two bad years in nine. His losses in those two years must be charged against his profits for seven years. Grower No. 2 did not have to gamble, because his acre costs were low. Even with a 50-bushel yield, he would not have lost much, although his income for his own labor would be low. Grower No. 1 actually would have lost the cost of his fertilizer. The tendency in corn research work is to add materials and labor costs without considering the possibility of a profit. There are many considerations which can have a profound bearing on the success of a farm enterprise. In 1953 Grower No. 2 was broke and had no credit. In 1958 his labor income on his 176 acres was $67 an acre. At the end of his seven years, he owned a car, a pickup truck, farm equipment, and had $5,000 in a savings account. He was farming this family farm for half the crop. If he could do this, any farmer in the United States could do the same if he had the same mental make-up.

Grower No. 2 tried many things that might affect his yields. He found that subsoiling to a depth of 2 feet along with his liming program when the ground was very dry increased his yields 21 bushels an acre. Postponing his plowing and planting date from May 1 to the 15 increased his yield 35 bushels an acre. He believes in plowing and planting, thus reducing his fitting costs. He said that if you don't work the surface too fine you get better root growth and a minimum of weeds. He did not have to cultivate his corn because he did not pack the ground before planting. He claimed that cultivation reduced his yield by 9 bushels.

Thus, it is possible to make a profit growing corn if acre costs can be reduced and yields increased. Experiments such as those carried out by Grower No. 2 showed that land classified as submarginal and unfit for growing corn, probably considered worn out by many, can be made to grow big yields if attention is given to the critical factors. It was proven here that heavy applications of fertilizer did not increase yields.

The fertility level of the soil farmed by Grower No. 2 at the start of the program showed a low nitrogen level, 14 pounds of phosphoric acid, and 41 pounds of potash, with a pH of 6.8 in the acre-foot. At the end of eight years, after applying from 8 to 16 tons of limestone per acre and not over 40 pounds of 10–20–10 in solution per acre in any one year, the organic matter had increased from 1.4 to 2.1 per cent, the phosphoric acid to 41 pounds, and the potash to 77 pounds per acre. A test was made after a yield of 134 bushels of corn had been harvested and before the stalks were worked into the soil.

There is a general idea among crop research people that a 100-bushel corn crop so depletes the fertilizer that large quantities must be returned to maintain fertility. This has not been indicated in my experimentation on Ohio, Illinois, and New York State soils. It is one point on which I had to change my ideas radically after I finished college and came up against actual farm problems. Several experiments may be mentioned to show how the mental make-up of a farmer can have a tremendous bearing on his acre profits at the end of the year.

We decided in New Jersey that there were many things which were hurting farmers' yields. We decided to conduct a farm survey on 132 farms in New Jersey where market tomatoes were being grown. It costs considerably more to grow market tomatoes than canning tomatoes. The survey was conducted for three years. Each year the results were the same. We found that fertilizer had very little effect. Too much manure reduced yields, especially on high organic matter soils, because available calcium was too low. Aeration was quite important. The best yields came from the higher elevations and the lighter soils. The amount of limestone was correlated with higher yields, but costs were not correlated with anything. In very few cases did farmers make any profit beyond being paid for their labor. In other words, it was difficult to put your finger on any one thing that could account for a poor crop, except the lime content of the soil. The yield varied from 70 to 300 hampers per acre. Farmers had to pick over 230 hampers to break even on costs.

A friend of mine at the Michigan Experiment Station told me about a survey he made among raspberry growers. The yield per 1,000 square feet ranged from a few to many crates. I don't remember the number. The cost of growing, not harvesting the crop, varied from 37 cents to $2.32 a crate, and he could find no reason for so much difference. It could have been correlated with yield. He felt that the management of the beds had much to do with it.

From these observations, I have concluded that our food production problem is not simply a matter of dumping on fertilizer. As a matter of fact, the need for fertilizer probably will play a minor role in our problem of feeding future generations.

I do not say the plant food is not needed to grow 100 bushels of corn, but apparently, it is coming from minerals in the soil, and we can continue to grow that crop year after year if we maintain calcium saturation of the soil at the proper level. The plant food is made available through weather agencies every year, provided a satisfactory level of calcium is maintained. It makes sense to use this available plant food, because if it isn't used for crop production it probably will be lost by leaching or surface run off, and eventually it will feed the fishes in the Gulf of Mexico rather than a crop of corn in Ohio.

Countries growing insufficient food have given serious consideration to means of extending crop lands through irrigation and increasing yields by means of fertilizer. No startling increase in food has resulted, because in many cases the limiting factor or condition had been ignored. A proper evaluation of the reasons for low yields had not been made, so a remedy was not available. My opinion is that in every case, the first experiment should be exploratory. A series of plots should be initiated where varying amounts of pulverized limestone have been applied and thoroughly mixed with the soil. Fertilizer may be applied in cross strips, depending on the type of soil. A check plot should be included receiving no limestone and another receiving neither limestone nor fertilizer. In general, very little response will be seen from fertilizer the first few years, until the limestone has had a chance to become part of the colloidal complex.

I would establish the plots by covering one acre of ground with 1 to 5 tons, another with 6 to 10 tons, others with 11 to 15 and 16 to 20 tons of limestone per acre, if it is a heavy clay soil. In a sandy soil 0, 1, 2, 3, or 4 tons may be enough. On a silt loam 0, 2, 4, 6, and 8 tons should be applied. On a clay loam 0, 5, 10, 15, and 20 tons and on a muck or high organic matter soil 0, 10, 20, 30, and 40 tons per acre should be used in acre plots. By making the plots an acre in area, they will be large enough should it seem desirable later to superimpose fertilizer plots on the limestone plots.

In tropical and semitropical areas the soil acidity test is of little value in determining the lime needs of the soil. Since the calcium available to the growing crop is the important consideration, a calcium test should be developed and standardized against crop yields. The actual calcium needed to saturate the colloidal complex (clay and organic matter) must be determined for each soil in the area. The purity of the limestone, its calcium and magnesium content, its fineness and hardness should all enter into the calculations determining the limestone needed in an acre-foot of soil. If the soils are of acid origin, the calcium needed for 3 feet of soil must be determined. I usually determine how much is needed in an acre foot and then multiply by 4, which gives me the amount of limestone needed eventually to grow the maximum yield. On soils of limestone origin, the determination of calcium in the plowed layer (one foot deep) may be sufficient.

From these calculations, it may seem as though we are applying so much limestone that it would not be economical to grow a crop. The purpose is to saturate 85 per cent of the colloidal soil complex with calcium. When we have accomplished this, we should not need limestone again for ten or more years. Therefore, we don't charge the cost of the limestone against a crop in any one year. We can charge it against ten crops at least or consider it as part of the investment in land.

In temperate regions we have somewhat different conditions. The organic matter requires up to four times as much calcium to saturate it as a pound of clay. We have to take this into consideration in calculating the amount of limestone needed. Temperatures likewise must be considered. They have an effect on the speed of reactions, which in turn have a bearing on the accumulation of negative charges.

Very few of our soils, the world over, have come near the degree of calcium saturation of the colloidal complex necessary to get maximum yields when weather conditions permit. Because of the low yields due to inadequate calcium the cost of growing the crop exceeds the value of the crop at harvest.

In countries where labor is cheap enough, average yields adequately exceed the cost of production, just as they did in the early days of our farm operations. There, crop growing becomes more a means of livelihood than a bare existence.

Much can be done through research, when it is carried on by people with the proper point of view. Research for the sake of accumulating knowledge has its place, but research that will help the farmer to raise his standard of living is more popular and is more readily supported by public funds. To get results that really show how to get bigger yields requires the efforts of someone with experience. We have wasted a lot of research money supporting studies of people who lack training and experience. They have very little idea what it is all about. The proper unbiased point of view is very essential.

The earth's crust is well stocked with nutrients which will remain there until some scientist finds out how to adjust the chemical and physical conditions to release them for plant growth. We have paid too little attention to this phase of the problem and too much to the idea that if a soil does not produce a good crop, fertilizer probably is needed.

We also have large areas in Africa and South America waiting for a smart plowman to turn some furrows and reap a fortune. With the help of irrigation and the use of limestone, adapted varieties, and cheap labor, we can feed the world population for many years. I doubt whether fertilizers will be needed in appreciable quantities. Experiments, of course, should be initiated to determine whether appreciable quantities of fertilizer are necessary to produce top yields.

Experiments conducted in the past, without regard to the physical and chemical condition of the soil, have contributed very little factual information to our knowledge of fertilizers. With a broad, unbiased approach, we can hope for much higher production levels at much lower costs. When Malthus set forth his doctrine on world and food population, he did not reckon with the imagination of trained research men and smart farmers who are capable of reasoning out a possible solution from a collection of data. I have worked with farmers who do not have the advantage of a college education but are better researchers than some college-trained men.

Dry fertilizers, when applied in the soil, vary in availability to the plant. Available rainfall and soil moisture pretty well determine how much we can expect to get into the growing crop. The ingredients in dry fertilizers determine their availability. The condition of the soil with respect to lime, the amount of clay, and organic matter, have a bearing.

As a result of the insolubility of dry fertilizer in dry soil, I decided to try fertilizer solutions as early as 1931. I used a 5–10–5 dry fertilizer, dissolved what I could in water, and compared it with plots with the dry 5–10–5. It gave me answers to many of our fertilizer problems far different from what most agronomists are willing to admit. Unfortunately, the greater efficiency of the solution over the dry fertilizer was not attractive to fertilizer sales and research has not kept up with its use. It has so many possibilities that I expect to see its use increase in popularity; and, because of its efficiency, it will play the major role in the growing of crops in the future. The fact that it can be handled by pumps and pipes just as the other liquids are handled is the major factor in its adoption as a main source of plant nutrients. Also, it has been used as a foliage spray on all crops, and small quantities have increased yields appreciably.

CHAPTER 9

Drought and Rainfall Control Yields But There Is Much Man Can Do to Offset Their Hazards

WATER IS all-important in the growth of high yields. In desert areas, irrigation may control the growth of crops. Without rainfall, we may grow fair crops. But where we depend on rainfall, we can be hurt by too little rain and by too much rain. In itself, rainfall is not the determining factor. What we do with it when it falls on our land is the important consideration. This determines how much damage insufficient or excessive rainfall does. In most cases, our utter dependence on weekly or biweekly rainfall is due to poor farm management.

Distribution of rainfall affects most of our practices. A widespread lack of rainfall can do us much damage if we let any rainfall run off. With proper management, every drop of rainfall should be absorbed by the soil with little surface erosion and stored in the lower levels, where the roots can reach it when they need it. This could make the difference between no crop and a good crop. I have seen it make the difference between 25 and 125 bushels of corn. The problem of getting the water to soak in is discussed in the chapters on subsoiling and liming.

Tomatoes ordinarily grow best on a moderately dry soil. I have seen a 25-ton yield produced on a soil that had one rain of one inch two weeks after the plants were set, and no more rain until the crop had been harvested.

Too much rain may or may not damage a crop. If we have excessive rain and the water runs off or infiltrates the subsoil, it usually does little damage. When stored in the subsoil for future use, it is ideal. If the water soaks into the plowed layer and stays there because of a plow sole, roots can be smothered, especially if it happens during periods of hot weather—about 70 degrees or above. An increase in temperature throughout the growing range for a given crop speeds up respiration, which means more rapid exchange of carbon dioxide and oxygen. Any interference with the removal of carbon dioxide from the root surface may cause degradation and death of the cells which are sloughed off by the plant. If the temperature stays cool, very little damage may result, because respiration is slowed down. At 70 degrees or higher temperatures, the plant becomes very active—respiration is rapid and the roots must get rid of carbon dioxide before it becomes toxic. If it stays around the roots, they die from lack of oxygen. The roots must absorb oxygen from the air in the soil, and if the soil is full of water, there is too little change of air. The only cure for this is to have soil open enough so the water can leach down to the natural water table.

Most thinking on this subject is directed at tile drainage, but that thinking is faulty. Tile drains are only desirable for the purpose of lowering the water table. For this purpose, tile has a very limited use, since there are very few cases where it is desirable to lower the water table. If we tile all our soils, all we do is allow the rainfall to trickle or seep through the soil, collect calcium and fertilizer nutrients on the way down, and carry them to the tile so they can be carried off the land to the rivers and the ocean. A soil with natural drainage should not be tiled. If it stays wet, the physical condition must be changed by applying limestone. The only advantage of carrying water away from the land is to feed the fish. I wonder how farmers can afford to do this? They buy fertilizer to put on the land and then tile it, so the soluble part of the fertilizer, which our crops need, runs to the rivers and eventually to salt waters.

Rainfall causes floods if it is excessive. The only measures we have taken to control floods are to build dams to hold the water in large reservoirs to slow its race to the ocean. We complain that farmers have cut down the forest so the water flows off too freely. Few seem to realize that our river valleys were formed by floods long before man was on this earth. Heavy rainfall has always caused floods. Damming rivers to hold back water is a worthwhile procedure if we use the water for irrigation when we need it. Floods cost farmers money, not so much because of physical damage, but because so much good soil and plant food is carried away.

If we would subsoil and apply sufficient limestone to our cultivated land so that the soil could absorb the water as fast as it fell, we would not only control our floods, or at least greatly reduce water runoff, but we would more than double our yields —and there is also a possibility that we would reduce the money spent for fertilizer.

Excessive rainfall, properly handled, will do minimum damage, and can do much good if stored in the soil for future use. We must know soils to know procedures. The big problem is to get the water away from the roots in the plowed area and to encourage roots to seek water in the subsoil. In this way, the surface soil is free of excess water and it is possible for the roots to get all the air they need. This is best done by providing good aeration with a subsoiler and adequate applications of limestone. This results in good surface drainage. And when I say good drainage, I mean free seepage from the surface to depths of 2 to 3 feet in the subsoil.

A farmer in one of the eastern states came to Mr. Charles Nissley and myself about several ponds on his farm. He was renting the land and had the option to buy the farm at a low price, which he would do if we could show him how to drain it. One 25-acre field on which he usually grew potatoes had a 2-acre pond in one corner. It had been used for a skating rink every winter for fifteen to twenty years. Every time it rained, the water ran off the field to the pond, keeping it filled. There was no outlet nor could the water soak into the subsoil, so the pond always had water standing in it.

The first thing we did was to run a subsoiler 25 inches deep, at 3-foot intervals around two sides of the farm side of the pond. The other two sides had a fence row and a state highway for boundaries. Following the subsoiling, the pond dried up, because rainwater could not run into it. Then we limed the field with 4 tons of limestone and subsoiled it lengthwise and crosswise to speed up the movement of the limestone into the subsoil. All the fields had the same condition—temporary ponds had formed in each. We gave them all the same treatment, with the same results. A permanent pasture, which was very rough, was also very swampy—this, too, was given the lime-subsoiling treatment. It dried up the swamp and permitted the farmer to establish good pasture for night use. He grew good corn and potato crops every year after that, because he stored his rainfall in the subsoil, where it could be used by his crops in case of drought.

Almost every year in some area of Ohio as well as other states, we have rather heavy rainfall in May and June. In those areas, particularly in unlimed soils, corn comes up and, when it is a foot high, turns yellow—in spite of the fact that the farmer followed state recommendations. Some of these farmers side-dressed with nitrogen, but it did no good because the cause was not a shortage of nitrogen. The real problem was a deficiency of calcium, which prevented good physical structure and prevented water from moving away from the roots to lower depths. The stagnant water soon lost its oxygen and the corn roots smothered. If the ground was dry enough, you could cultivate; but when it dried, the corn would turn green anyway, if root growth was still possible, because when the water left, air immediately penetrated and supplied the necessary oxygen.

A number of years ago, in one eastern state, we had heavy rains when potatoes were 8 inches high. The foliage in the fields turned yellow, and, thinking that the nitrogen had washed out of the soil, the fertilizer people sold a lot of nitrogen for side-dressing. I was working with Mr. Fred Bateman (*Farquhar* and *Iron Age*) on fertilizer experiments. He had a 50-acre potato field which was very sandy and slightly rolling. His potatoes turned yellow in the lower part of the field. We walked between the rows, and noticed that every fifth or sixth plant was green. Those in between were yellow. We got a spade and dug up 10 feet of one row. The green plants had their roots well into the subsoil, 16 inches deep. Those between had their roots only in plowed soil.

When we checked further, we found the green plants lined up across the field in rows 5 feet apart. I suggested it was a carry-over effect from fertilizer the previous year. "Last year," he told me, "the field was in rye with no fertilizer. The year before it was in corn, but the rows were fertilized the same direction the potatoes were. The year before that the field was in tomatoes. They were planted in rows crosswise to the potatoes and five feet apart." "What did you do to the tomatoes?" I asked him. He subsoiled under the rows and put a mixture of one-quarter gypsum and three-quarters limestone under the plants. He made a furrow a foot deep, placed the mixture in the bottom, covered it up, and set the plants on top. "We had a big crop of tomatoes," he told me. The gypsum had promoted deeper drainage and aeration.

Here again we have a demonstration of the damage that too much rainfall can do if the water can't seep into the subsoil because of a plow sole. In a dry season, or one with normal rainfall, you might not see any effect, except that the yield might be much better where the plow sole was broken up.

It has been my experience that we have fewer problems with heavy rainfall on sandy loam soils with low levels of clay and organic matter than we do on those heavy soils where clay and organic matter are high. Water moves more slowly through the heavy soils and, during hot weather, damage from heavy rain can occur before the water has a chance to seep below the plowed layer. With continuously heavy rainfall, the plowed layer actually becomes swampy and we have surface swamping on what is considered highly fertile soil. We must not underestimate the oxygen problem here. Adequate limestone plays an important part in correcting it.

In areas of heavy rainfall these heavy clay soils are ridged up and crops are planted on the ridges. The first time I saw a field of back furrows on 40-inch centers, I thought the person was out of his head, but when I saw nice green corn in spite of a wet spring (while a neighbor's corn was yellow), I realized that this was a practice assuring good aeration. When I talked with the farmer and asked him why he did it, he said that his father always did it in coastal North Carolina. Since then my experience over the years has shown that it is an easy way to get good yields on soil that is high in organic matter, high in clay, and where water seepage horizontally or vertically is too slow in hot weather when respiration in the plant is at its maximum.

A farmer called me and wondered why corn on one side of his field was not growing satisfactorily. When I saw the field, I realized what had happened. He did too, as soon as I started to tell him. He had a field of sugar beets alongside the corn. They were growing very well. He had prepared the sugar beet ground a month before the corn ground was prepared. He overlapped on the corn field and several weeks later, when he prepared his corn ground, this strip was worked a second time. The heavy rains came when the corn was a foot high. Most of the corn recovered and made a good growth, but the strip that was "overworked" never entirely recovered.

The year 1960 was a wet season in parts of Ohio. We have an experimental farm at Olena. It is in an area of silt and clay loam in which most of the rainfall runs off the land. It is low in calcium but has a fairly high pH. We had heavy rainfall in May and June. The soils on neighboring farms were worked excessively, because every time it rained it had to be worked again. This happened three times on some farms.

Corn was planted on these fields and it germinated well, except in areas that were under water. Since soils on the Olena farm were so variable, we had spotty fields. We plowed and planted on our wet soils and grew over 100 bushels of corn. But with several more rains the corn on much of neighboring land turned yellow and stayed yellow all season, in spite of the fact that some growers side-dressed with nitrogen fertilizer. The yield naturally was very poor. On our farm we plowed late and planted corn after June first. We did not work much of the ground, had no yellow corn, and harvested as much as 137 bushels. The rains did us a lot of good, because our ground was limed and was not packed from overworking. Where we had a low calcium reading and considerable clay and organic matter, our corn turned yellow, because the roots could not get oxygen. We applied oxygen around a few plants in a test plot and grew 100 bushels of corn instead of the 35 bushels where we did not apply oxygen.

We had some yellow corn that was planted earlier on ground that was disc-harrowed once. When the corn was 15 inches tall, we pulled the subsoiler between the planted rows to get some air down to the roots. It was amazing how soon the corn turned green and started to grow. When we harvested the corn, we found that subsoiling had increased our yield from 50 to 103 bushels.

In this same area, but where the corn was hurt the most, we selected six plants in six locations in the area. Plants 3 and 4 were definitely poorer than 1 and 2 and 5 and 6. We bored two holes 12 inches deep alongside each plant. The holes around 3 and 4 were filled with pure oxygen gas and sealed over. All the holes were sealed over. From Plants 3 and 4 we harvested ears that were 8 inches long and well filled out. The ears on the other plants were poor nubbins 2 to 4 inches long.

Too often we think that yellow corn means nitrogen deficiency, that purple corn means phosphorus deficiency, and that marginal burning of the leaves means potash deficiency. Perhaps they do; but a lack of calcium, too much rainfall, or compacted soil may cause all of these characteristics to appear. These are deficiency symptoms, but they cannot be corrected merely by supplying the deficient ion. This confuses the issue when we try to correlate soil tests with yields. Because of the variation in rainfall, each farm in each area can have different results and different problems from one year to another. One wishes that a farm could be run with the aid of a slide rule or a chemical test. Such a utopia is a long way off. Perhaps we will have to wait until we can visit other planets to find the answer. So far our brains have not even assured us of consistently high yields on one given farm in one particular area.

Many cultural practices play an important part in our yields. As colloidal clay and organic matter increase in quantity in a soil, our cultural problems become more complex. Farmers have practiced subsoiling for many years, but most experiment stations that have investigated the practice at intervals have concluded that the practice has no value. Conducting such experiments without regard to the water content of the soil is a waste of time. The object in subsoiling the ground is to break up the subsoil, leaving fissures running in all directions. This promotes better movement of air and permits rapid movement of water from the surface to the subsoil, where the water can be stored for future use. It also helps to speed the movement of limestone from the surface to greater depths, which encourages roots to penetrate the lower soil horizons.

We have also used the subsoiler to open the soil between rows of corn where the foliage was 16 inches tall and exhibited a yellow-green color indicating nitrogen deficiency. Three days after this practice was followed the corn turned dark green. The practice increased the yield from 47 bushels to 103 bushels.

Subsoiling has always given a worthwhile response when the practice was followed during the season when the ground was hard and dry. Many farmers who have silt and clay loam soils plow their ground when it is too wet. If the soil is wet enough to show a glaze after it is plowed, it probably is too wet. Farmers are anxious to plow their ground as early as possible, partly to get the work done. Often the surface of the soil is dry, but the bottom of the furrow is wet. As heavy equipment moves over the surface, the plow sole becomes puddled, and bakes hard as it dries out. Roots won't penetrate this hard subsoil. Thus the roots are shallow and are at the mercy of the weather, particularly when it is dry. Farmers often work the fields several times after plowing, and if rain should fall, they have to work the soil again. The soil becomes packed and weed seed germinates in abundance. Then it is necessary, by means of a rotary hoe and cultivators, to control weed growth and loosen the soil so that the roots of the crop have adequate air.

Minimum tillage as a practice of plowing and planting is often practiced for spring crops. The ground is not plowed until it is time to plant the crop. The ground remains loose, and because it does not provide sufficient moisture, weed seeds won't germinate until sufficient rain has fallen to pack the soil around the seed. Growers have told me that they have practically eliminated troublesome grasses by following this procedure for three years.

Minimum tillage can be practiced on those soils that are well supplied with calcium. Soils that plow lumpy are not ready for minimum tillage. A combination of minimum tillage and subsoiling makes it possible to plant crops closer together, which in the case of corn may account for an additional 40 bushels.

Our records show that with certain practices on a soil that has 85 per cent calcium saturation, you can expect the following yield increases:

1. Soil which needs 1,600 pounds available calcium	67 bu.
2. 1,600 pounds calcium applied as limestone	123
3. Yield increase due to more stalks per acre 12,000 to 17,000 on No 2	153
4. Number 3 subsoiled previous August	166
5. 2 gallons 10–20–10 on the seed and 2 gallons 10–20–10 with nutritional additive as foliage spray	199
6. Minimum tillage: plowed and planted	208

Very often, if a farmer's soil needs limestone (low calcium reading) and he is unable to apply all the limestone he really needs, it is possible to apply 300 pounds of a finely ground high calcium limestone to the row in a 4-inch-wide band over the seed. Corn growers plant the seed with 2 to 3 gallons of 10–20–10 solution, fill the fertilizer hoppers with calcium limestone, and allow the limestone to drop over the row on the top of the ground.

The following experiment was initiated on a piece of ground that showed an available calcium reading of 400 pounds. It needed 2,800 pounds to be properly limed. The results were rather surprising.

1. Corn planted with nothing added	40 bu.
2. Corn planted with 2 gallons 10–20–10 fertilizer solution on seed	54
3. 300 pounds high calcium limestone spread along row over seed	68
4. 2 and 3 combined	103

I recommend this program where a farmer cannot apply his limestone broadcast.

FIGURE 3

YIELDS OF WHEAT OBTAINED ON LIMED GROUND

Varying amounts of 7–14–7 fertilizer solution applied broadcast at seeding time

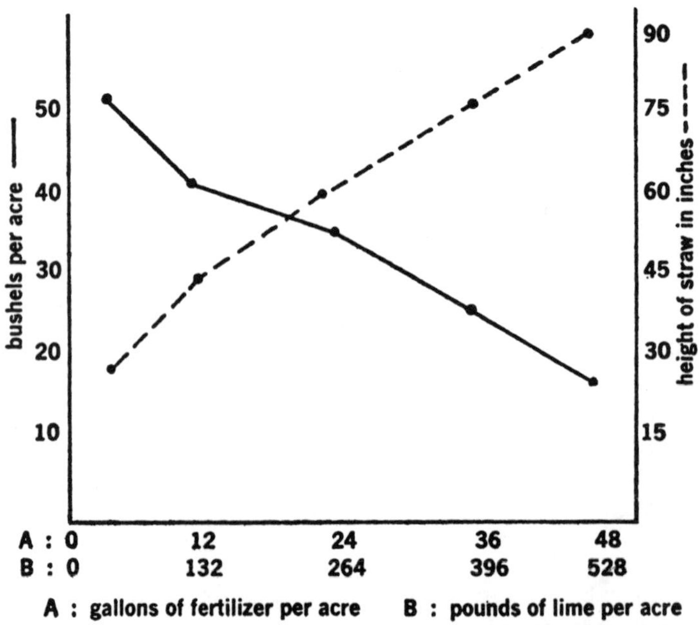

The results in Figure 3 are typical of all areas where we apply as much solution as we would dry fertilizer. Leaves and stems are grown at the expense of grain as we increase the amount of fertilizer.

FIGURE 4

COMPARISON OF LIQUID AND DRY 5–10–5 FERTILIZER FOR SWEET POTATO YIELDS

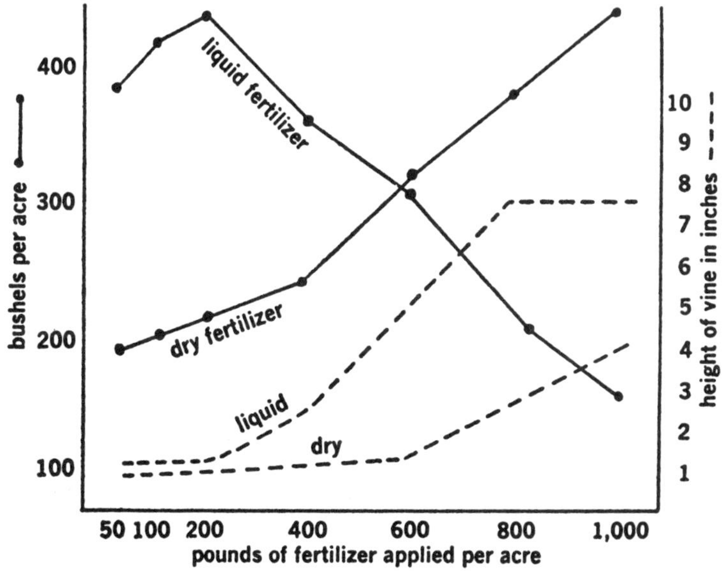

Sweet potatoes were grown with varying amounts of fertilizer in dry and solution form. Figure 4 represents the yield of choice tubers and top growth with varying increments of dry and solution 5–10–5 fertilizer. Making a solution out of the 5–10–5 greatly increases the efficiency to the plant in using it to promote growth of foliage. The yield, which depends on how efficiently the plant can manufacture starch and sugar, depends on the direction that the plant is expending its energy.

These potatoes were harvested on October 1. Potatoes with voluminous tops will increase their yield of tubers by 100 per cent if left in the ground until frost. As temperatures drop, more starch is stored and growth of vines ceases, in spite of luxuriant top growth.

CHAPTER 10

Fertilizer Solutions Have Many Advantages Over Dry Fertilizer

THE RATE of intake and utilization of dry fertilizer is greatly facilitated by dissolving the soluble part of the fertilizer in water. The efficiency of utilization in promoting higher yields of either dry or solution fertilizer depends almost entirely on how closely the calcium saturation of the base exchange approaches 85 per cent of the total.

In the early days of fertilizer solution research, solution was nothing more than a given dry fertilizer placed in a tank filled with water. By intermittent stirring over a two-day period, permitting the insoluble material to settle out, we eventually obtained a clear solution that had a 1–1–1 ratio of nitrogen, phosphoric acid and potash. If we dissolve a 5–10–5 fertilizer in water, we end up with a 2.5–5.0–2.5 solution, if we use equal quantities of water and mixed fertilizer. When I used this in the field, I found that 500 pounds of this solution was far more efficient than 500 pounds of dry fertilizer. I had a much greater volume of leaves and stalks, but the yield of seed and tubers was not as large as when I used dry fertilizer.

It brought up the question as to the amount of fertilizer to use to get as good a yield with this solution as with dry fertilizer. I set up an experiment, using sweet potatoes as my test crop, where I made comparisons of solution and dry using a 5–10–5 fertilizer. Each succeeding plot received 50 pounds more than the previous plot until I had twenty plots of both solution and dry. The results have been discussed elsewhere. Briefly, I found a tenth as much fertilizer in solution grew as good a yield as dry fertilizer did. This relationship still exists and can be proven with field plots on adequately limed soils.

The mechanics of making these solutions dictated many changes in future research. I began to look for salts which were practically all soluble in water and began to use urea, ammonium phosphate, potassium nitrate, potassium chloride, ammonium nitrate, and potassium phosphate, and found that I could make a 7–14–7 solution. But this mixture was toxic if used as a foliage spray in concentrated form.

In order to find a material that could be mixed and applied to the foliage, I finally went to materials that were safe to use in concentrated form. By eliminating nitrates and chlorides, I was able to make a 10–20–10 fertilizer that was close to neutral, that was non-corrosive to metals, and was not toxic to foliage when applied in small quantities. Because of the high cost of ingredients, some of which were dangerous to handle, what I made, most mixers would call too high-priced. However, if we consider how little is actually needed to grow a crop, it drops the acre cost for fertilizer considerably below the cost of dry fertilizer. And since these characteristics can be demonstrated, it is a good indication that the use of fertilizer solutions for ground application and foliage applications is a worthwhile contribution to the efficient production of food.

The application of bulky liquids where as much plant food is applied in solution as one would apply in the dry form is not based on good research and, because it offers no advantage over dry fertilizer, probably will not contribute to the use of fertilizer in the production of crops. To be worthwhile a fertilizer solution must be efficient, non-corrosive, non-toxic, and non-poisonous to humans. It must not add to the labor needed to apply it. It must be competitive with dry fertilizer in acre cost. Experiment station personnel insist that if you need 500 pounds of dry fertilizer, you must apply the same amount in solution. My first experiments showed this to be faulty reasoning that cannot be supported by research investigations.

When I first compared the two (dry 5–10–5 with the same amount of the 5–10–5 in solution) on tomatoes, sweet corn, peppers, and lima beans, my results were very disappointing. I found the growth of the plants was dark green and lush in appearance, but the yield of fruit or seed was very much more where I used the dry fertilizer. I realized that I was using too much fertilizer. Apparently, by dissolving what I could of the dry fertilizer, which was a 3–3–3, after several days of soaking and stirring, I was making the fertilizer much more available to the growing plant. I soon found that 20 pounds of this dry 5–10–5, when dissolved in water, did a better job than 100 pounds did when applied in the dry form. Also, the efficiency was stepped up with other practices which made solutions more and more attractive to use.

This type of solution is still being used and recommended by experiment station personnel on a pound-for-pound basis. It is referred to as a "bulk liquid," but has few if any advantages over dry fertilizer and has some disadvantages. It is not backed by good research studies. It is in a class with nitrogen solutions and anhydrous ammonia. It is being used too liberally, and except for leafy growth of poor quality, it has less and less use in solving our fertilizer problem.

I found that if fertilizer solutions were to be useful, it would be necessary to use them as seed treatments (suggested by Dr. Roberts at the Rothamstad Experiment Station in England), as foliage sprays (as recommended by the horticultural group at the Michigan State Experiment Station and the North Carolina Experiment Station some ten years ago) and that many refinements would have to be made.

The 10–20–10 was the result of many years of research. It is needed only in very small quantities to get better results than I could get with dry fertilizer. This 10–20–10 also had all the trace elements that might be needed by the growing plants. It remained for comparison field plot tests to prove its worth, and after ten years of field study, it has met all the qualifications necessary for this type of material.

There is much confusion in our thinking about the use of fertilizer solutions. In the following discussion, I am referring to the 10–20–10 mixture as I have formulated it unless I state otherwise. The material is safe for animal and human consumption and carries no hazards if children accidentally drink it. It does not evaporate from the soil. In an open container a little water may evaporate from it, but its specific gravity prevents any appreciable loss of water. I have worked with these fertilizer solutions since 1925. Much research associated with this 10–20–10 was the result of my studies.

There was very little known about fertilizer solutions when I started to study the possibilities. At the present time, very few agricultural experiment stations have projects concerning the use of fertilizer in solutions. The reason for this is that the brakes were applied by the fertilizer industry and its propaganda organization, the National Plant Food Council. They have stood in their own light by trying to increase volume sales at all costs. I am afraid that now a few who have pulled their heads out of the sand are beginning to realize that the costs are higher than they can bear.

Fertilizer solutions are merely a means of helping the farmer get more good out of every dollar he spends for fertilizer. They are the result of a better understanding of the chemical workings of crop plants and soils, of what our soils can support in the way of crop growth, and of how much a farmer must add to his soil to make the most profit per acre. Because they are 100 per cent available to the crop, they are needed in only small quantities. One hundred pounds of a 10–20–10 mixture with trace elements in proper combination will do a crop more good than 600 pounds of 5–10–5 dry fertilizer.

When Russell Brothers of Milan, Ohio, came to me with figures showing that they had doubled a 57-bushel corn crop by spraying the foliage with 20 pounds of 10–20–10 in solution two weeks before the tassels showed, I did not doubt their word, because their neighbor had told me that he had harvested 135 more hampers of sweet corn as a result of a foliage spray. These results seem phenomenal if we think in terms of pounds of nitrogen, phosphoric acid, and potash, and yet farmers have reported such results with many crops. After eight years of use, 90 per cent of the growers have reported profitable returns from foliage sprays.

For twenty-five years, interest in the application of fertilizer dissolved in water has been mounting, along with the usual praise and condemnation that accompanies any new product or practice. There was a time when many believed that the automobile was not practical, the airplane would never fly, and fertilizer, as a substitute for manure, would poison the soil. We have some people who won't use dry fertilizer today, because they say it doesn't pay to use it.

Atomic energy research by horticulturists in several agriculture experiment stations has opened a wide vista of possibilities in the use of chemical fertilizers and has encouraged interest in fertilizer solutions. It has laid the ground work for more efficient methods of supplying plant nutrients to growing plants. (I am using the term "fertilizer solution" for complete mixtures which contain both major and minor plant nutrients. They may be applied to the soil or, if properly compounded, to the foliage of plants.)

The greater effectiveness in utilization of plant nutrients from solutions under variable weather conditions, the elimination of waste in fertilization practices, and the elimination of hard work on the farm all focus attention on the use of fertilizer solutions and all will contribute to reducing costs of growing various crops.

Fixation problems occur in different soils when dry fertilizer is applied. The calcium base saturation of the soil becomes inadequate or too high—as in volcanic soils in which sodium, potassium, and hydrogen ions take the place of calcium. In such cases, foliage sprays have been demonstrated to give favorable results. Whether we apply fertilizers in dry or in solution form, calcium unsaturation can be a limiting factor in growth response. Plants are unable to get what they need through their roots. Many tons of fertilizer are wasted every year because we do not apply sufficient lime. Many soils that test neutral do not have sufficient calcium and magnesium. Our problem is to properly balance the ions in the soil. If calcium and magnesium are not adequately supplied, other elements like potassium, manganese, and trace elements won't do any good.

As our knowledge of plant nutrition increases, we can expect the application of foliage-applied nutrients to become more important in our fertilization program. At one time, we were all certain that mineral nutrients could not be absorbed even though gases could move freely in and out of the leaves. As the volume of research increases on foliar applications of insecticides, fungicides, weed killers, mineral ion sprays (particularly the trace elements), and exposure to injurious gases, we realize that we must modernize our thinking. The absorption of materials by way of the foliage, however, is not always a sure-fire method. The plant has no mechanism by which it can absorb ions outside of simple diffusion in a film of moisture. Thus, a wet leaf will take in nutrients, whereas a dry leaf will not. Applying nutrients to a leaf covered with dew assures its entrance into the leaf. Environmental conditions, therefore, determine the efficiency of absorption of foliage-applied materials. I have found that all plants respond in a similar fashion, except that time of application may be different on annuals (1 year) than on perennials (2 or 3 years). Time of day can be a determining factor. Grain crops with a determinate type of flowering may respond differently from indeterminate types like tomatoes or melons, which will grow indefinitely if given a chance. Even though many of the experiments I mention have to do with corn and wheat, in principal, their response is no different from tree fruits or flowering plants.

When the cells of the leaf are full of water and dew collects on the foliage, any soluble material applied to the foliage has a good chance of diffusing into the cell and the leaf with considerable speed. However, if the foliage is dry and water vapor is leaving the cells rapidly, there is little chance of foliage-applied materials being absorbed. The moisture evaporates quickly and the dry salts are easily removed by air movement. For this reason, the application of foliage sprays should be made at times when the humidity is high—between 6 P.M. and 8 A.M., or on cloudy, drizzly days. This may interfere somewhat with airplane applications, because of visibility, but applications made during a hot, dry day with air moving may be worthless. Wetting agents tend to improve the response to foliage sprays. Usually several small applications of foliage sprays are more effective than one heavy application. Our results on many foliar applications show that with the application of 20 pounds of 10–20–10 per acre, we can expect to get increases of 10 to 20 per cent in yield on grain crops and even greater increases on peppers and some crops like cucumbers, cantaloupes, watermelons, pumpkins, and squash. During this past summer, one of our corn growers reported a 100 per cent increase in yield. This is the second time this has happened in my experience. Minor element deficiencies are easily corrected by foliage spraying under such conditions.

Chester Long, at Wild Rose, Wisconsin, applied 2 gallons of spray on pepper plants when small peppers were forming and increased his yield by 800 baskets per acre. To a person steeped in the idea that fertilizers determine the yield of a crop, foliage sprays seem too inadequate to produce results like that. Yet results like this have been obtained too often to be classed as coincidental. Foliage sprays applied to plants in a rapidly growing state may help to grow a larger plant with no additional increase in yield. Usually, if the growing plant is changing from a leaf growth to a fruiting growth, yield increases will result.

Many years ago, I planted corn in washed gravel (in a bench in the greenhouse) to which I had added some pulverized magnesium limestone. I planted corn seed 4 inches apart in rows 6 inches apart across the bed. Then I sprayed every two rows, leaving a check in three different places. On successive pairs of rows, I applied 2, 4, 8, 16, 24, 32, and 40 quarts of 10–20–10 per acre. The spray was measured out and applied undiluted with an atomizer. Each pair of rows was shielded above the gravel. No attempt was made to confine the roots nor prevent the spray from contaminating the gravel. The plants were harvested when the tallest were 30 inches tall. The check plants grew only 8 inches tall and were yellow. They apparently had got no nutrients though they were in between the others with their roots intermingling. Two quarts produced no increase in growth, but the plants were slightly greener. From 2 quarts, the volume of growth increased in proportion to the material applied, up to 24 quarts (6 gallons), when the 30-inch level was reached. Beyond this, the volume of foliage increased but the height did not exceed 30 inches. The foliage sprays produced the growth. I repeated the experiment several times, but did not always get the same response. (Should anyone want to try this on corn, I would warn them that corn planted in the fall quickly goes into the seed-producing stage and may not respond. When watering, the water must be kept off the foliage after spraying. Apply the spray when the humidity is high. Do not apply the spray when the air is hot and dry and the moisture is evaporating rapidly. Apply the spray undiluted. Too much dilution in 6- to 10-gallon applications may cause burning.) I have repeated this experiment several times and I have grown corn to maturity with only foliage feeding. Even with 40 quarts—10 gallons—of undiluted spray, there was no injury to the foliage. I have diluted 10 gallons to 100 gallons, with water, and got considerable leaf injury. When these solutions are diluted, large drops form on the foliage, which may burn.

I have grown cotton plants to maturity in soil in 8-inch pots with only foliage sprays. These plants had 64 mature bolls of cotton. The check plants had 7. All nutrients came through the leaves.

Let's go back to the time when fertilizer solutions were first studied at Rutgers by the Vegetable Crops Department. At that time some sensational results were obtained, although we should not attribute any miracle-producing properties to fertilizer solutions.

Mr. Isaac Harrison, a very good co-operator at Crosswicks, probably was the first farmer in New Jersey to use any appreciable quantity of fertilizer in solution. He dissolved dry fertilizer in water in a large cistern and pumped it into a tank truck. From there it was pumped to tanks mounted on his tractors and applied at the first cultivation to snap beans, with very favorable results. Much of our early research work was done on farms of this type. The late Mr. Fred H. Bateman of Grenloch, New Jersey, was a very interested co-operator. He grew 600 bushels of Chippawa potatoes and 460 bushels of sweet potatoes on his sandy loam with only solution side-dressings. Much of the preliminary work on fertilizer solutions was conducted in the vicinity of Grenloch before 1945.

The fact that fertilizer solutions are more effective than dry fertilizer is no condemnation of dry fertilizer, nor is it any commendation of fertilizer solutions. The point in question is our climate. We talk about dry fertilizer being soluble and available. True; but it is no assurance that it is going to be available to our crops every year. We have figures which show 10 to 25 per cent availability. In dry years, 10 per cent is high for dry fertilizer.

Water is required to dissolve dry fertilizer, and many times there is not enough in the soil. Some of our research people say you may only get 5 per cent recovery from dry fertilizer. Some say 10 per cent. Some say that if you get 25 per cent recovery, you are doing very well. Why, then, is there anything phenomenal when we get the same results when we apply only 10 per cent of the same fertilizer in solution form and get equally good yields?

We have formulated our ideas about the need of fertilizer to grow a crop from field results where we use large quantities of fertilizer, in many cases under conditions where lime may have been a limiting factor. It is difficult for me to state the certain amount of plant food needed in order to produce a given unit of crop. I have no argument with the idea that our crops, when analyzed, show a certain amount of phosphoric acid and potash, which, when divided by the bushels per acre, gives a certain amount of fertilizer needed to produce a bushel of a given crop. But this type of data does not prove anything. Indeed, if one were to scan through the literature, one would have trouble finding proof for such statements.

To me, fertilizers are only part of the story. They make it possible for the plant to produce fruit, seed, and storage products from gases in the air. We must first supply sufficient limestone to bring that soil to its highest level of efficiency. Then fertilizer will go to work for us. Whatever plant food becomes available will produce a certain yield. It is our problem to add enough additional plant food to get the maximum yield under the prevailing conditions. If we don't use the plant food made available in the soil, every year we will lose it in drainage or runoff water. Furthermore, we don't want the fertilizer to interfere in any way with seed germination and subsequent growth of the seedling. Any interference at this stage can reduce the ultimate yield. Much of our research on fertilizer placement has done more to eliminate the toxic factor than it has to give us better utilization of plant nutrients.

The amount of fertilizer that we must add in any one year to get a maximum yield may be one pound or 100 or more pounds per acre. The question in my mind is why we should dump 500 to 3,000 pounds of fertilizer on an acre when we need only 25. We are spending money for something that does not pay dividends, and farmers can't do that and maintain the standard of living to which they are entitled. Farmers don't do enough testing on their own farms. They depend for directions on agricultural experiment stations which already have too much to do to try to meet all individual problems.

In a field like agriculture (growing crops on many different soils), where we have so many variables, it is difficult to prove anything. In comparing fertilizer solutions with dry fertilizers, I find it is always safer to rely on many experiments in different locations. It gives you averages which are more convincing than individual costs. I have had positive results with fertilizer solutions in approximately 90 per cent of all cases, whereas, with dry fertilizer plots in the same fields, I have not had 50 per cent of the trials show a positive response. With fertilizer solutions, 10 per cent of the cases showed no response, but in those same cases, dry fertilizer showed no response either.

I have tried to find out why we get such responses. In most cases, unsuitable levels of calcium are responsible. It is difficult to convince farmers, but after the first year they become very alert. One grower near West Liberty, Ohio, did not want to leave a plot without dry fertilizer because he said he would lose so much money. When I told him that I would pay him for his loss, he went along with our experiment. When we harvested his crop, he had 123 bushels of corn without any fertilizer in the row, 97 bushels with 300 pounds of dry fertilizer in the row, and 131 bushels with 2 gallons of fertilizer solution in the row. He used liquid fertilizer solution in all his plantings for several years. He was finally convinced by a dry-fertilizer salesman that to use only fertilizer solution would wear out his ground. Now he is still trying to grow 100 bushels of corn, as he did with my program.

During the past five years, I have concluded experiments on at least seventy-five farms in different parts of Ohio where comparisons have been made among the recommended dry fertilizers, fertilizer solutions in the soil, and foliage sprays. I have yet to get figures which show that it pays to use dry fertilizers and I have yet to see a single case where 300 to 400 pounds of dry fertilizer has given results as good as 20 pounds of fertilizer solutions gave. The reason is that we know so much more about fertilizer in solution than in dry form under our variable weather conditions. The year 1956 was more favorable to the use of dry fertilizer but, from experiments harvested, there were no differences. We had such an experiment on the farm of Les Wildermuth near Canal Winchester, Ohio. When we showed a 10-bushel decrease of his corn, the extension people asked to make a check. When they had finished, they showed a greater decrease for dry fertilizer. This should not be any criticism of dry fertilizer. It is a criticism of the manner in which it is used. There has been a tendency to recommend more and more fertilizer. The advantage of fertilizer solutions is that, because of availability, they are used in small quantities.

When we step up quantities of fertilizer solutions, we may get injury to seed germination, or we may promote too much leafy growth. Actually, on the basis of results that we have obtained so far, it seems that the amount of fertilizer we use in solutions is approximately the same as what becomes available from 300 pounds of dry fertilizer. So even though we think that there is a large discrepancy between the amount of dry and solution fertilizer applied, the amount the plants get is actually the same. We are only supplying the needs of the crop. The fact that we apply large quantities of dry fertilizer does not mean that it is needed or used.

During the past ten years there has been much condemnation of fertilizer solutions by agencies that have had no experience with them. Needless to say, there has been much praise for fertilizer solutions by people who have used them. A few agricultural experiment stations are doing research work on the supplemental value of fertilizer solution and, in some cases, studies are under way to actually compare dry and solution forms of fertilizer.

When we first started work on fertilizer solutions at Rutgers, many dire predictions were made. They have not materialized. We know very little about the application of either dry or solution nutrients to our crops. We can't judge the value of plant nutrients in water by what we think we know about dry fertilizer. We have assumed that if we need 500 pounds of dry fertilizer to grow a crop, we must use 500 pounds of fertilizer solution. This statement cannot be proven at this time. We are comparing a 100 per cent soluble material with one that is probably only 25 per cent available.

Many of our soluble fertilizers, like 15–30–15, 11–24–11, 10–20–10, and 12–12–12, have been condemned in the past because of experimental comparisons made on a pound-for-pound basis. Dry solubles are as good as any other fertilizer, providing we consider their characteristics—including solubilities—and use them accordingly. Fertilizer solutions have many advantages, most important of which is that their use fits in with present-day plans to eliminate hard work on the farm. Fertilizer solutions are made up in liquid forms comparable to dry fertilizers. That is, 100 pounds of 5–10–5 fertilizer solution would contain the same nutrients as 100 pounds of dry 5–10–5 fertilizer. While the solution can be pumped from one container to another, the dry fertilizer must be lifted. Here in Ohio, fertilizer solution is handled just as you handle gasoline—tank truck to farm tank to a tank on the tractor or other equipment.

Growers in Ohio fill up their tractor tank in the morning and plant 15 to 20 acres of corn without touching a pound of fertilizer. As one grower told me, "Since I use fertilizer solutions, I feel like going to a dance after planting corn all day. Could you imagine that, if I handled fertilizer bags all day?"

I asked one grower how much fertilizer solution he used. He said, "Two gallons of 10–20–10 in the row when I plant the corn, and then I apply two gallons per acre on the foliage."

"How do you apply the fertilizer on the foliage?" I asked.

"I have a flyer come in about two weeks before the corn is ready to show tassels and spray two gallons per acre without diluting it. He can fertilize twenty acres each time he goes up with a load."

"In other words, you apply forty pounds of 10–20–10 to the acre and grow a crop of corn? Aren't you afraid of wearing out your soil with such a small amount of plant food?"

"That field"—he pointed to 40 acres west of the road—"has been in corn for nine years, eight years of which it has been fertilized with that amount of fertilizer solution. Does it look like worn-out soil?"

"No," I said. "It looks like one-hundred-bushel-to-the-acre corn. Have you always grown one hundred bushels of corn per acre on this soil?"

"While I was using dry fertilizer, according to recommendations, I never grew over seventy-five bushels. This year I have corn that will go one hundred and fifty bushels an acre. I can grow more corn now even with dry fertilizer, but why lift all of those bags when I can do as well or better with liquid?"

"Surely you don't attribute that all to liquid fertilizer?" I asked him.

"No," he said. "All this land has had at least eight tons of limestone per acre during the past eight years. Some had sixteen tons."

"How do you know that dry fertilizer would not increase your yields as well?" I queried.

"I have dry fertilizer check plots in every field for comparison, and I have yet to see as good a yield from four hundred pounds of dry fertilizer as I get from forty pounds in solution. Last year my dry fertilizer plot cost me money. I grew less corn than I did with no fertilizer. The solution made me a good profit above the cost."

I went to see another grower who has been on a solution program for three years. I asked him whether his soils were less productive as a result of using only liquid for three years. He told me that his yields were getting better every year. Both of these growers raise beef cattle and had some manure, which would help to keep their soils from "wearing out."

I talked with a grower who lives near Columbus, Ohio. He does not raise beef cattle but has a few sheep. He has been a co-operator of mine for seven years. Much of what we have learned about fertilizer solutions came from experiments on his farm. I asked him whether he thought he would wear out his soils by using comparatively small amounts of fertilizer in solution form.

"So far there is no indication of it. My crops are better now than they were when I started on this program. My yields have increased from sixty-seven to well over one hundred bushels per acre, and my highest yield was one hundred and ninety-one bushels. You must remember that this is not potentially fertile soil. People won't believe me when I tell them I grew one hundred and ninety-one bushels of corn on some of these rolling, gravelly hills, with fertilizer solutions." Most of these one hundred acres were used for some form of an experiment. He told me that if fertilizer solutions did not give him results there was no use trying to get results with dry fertilizers.

A brother team in southern Ohio had experiments on potatoes. They grow approximately 100 acres. The plots where they sprayed the foliage with 5 gallons of fertilizer solutions per acre, incorporated with their regular spray program, increased their yields from 567 to 637 bushels. They also found that by spraying the seed with 3 gallons of 10–20–10 as they planted it and 5 gallons per acre as a foliage spray, they could grow as many potatoes as they did with their dry fertilizer. In other words, at a cost of $20 an acre for fertilizer solutions, they grew as many potatoes as they did with $70 worth of dry fertilizer.

It is possible to use too much fertilizer in solution form. We had an experiment on wheat. This field was well limed before the wheat was sown. We applied 100, 200, 300, and 400 pounds of 7–14–7 broadcast in strips across the field before the wheat was sown. When we harvested the wheat, the plot with no fertilizer yielded 52 bushels; 100 pounds of 7–14–7 yielded 37; 200 pounds yielded 33; 300 pounds yielded 23; and 400 pounds yielded 17 bushels per acre. But the yield of straw increased with each increase in application of fertilizer solution. With 400 pounds of fertilizer, the straw was over 6 feet tall and heavily stooled. I realized that we overdid this, so on later experiments we used 20, 40, 60, and 100 pounds per acre. In these experiments, 20 and 40 pounds of 10–20–10 fertilizer solution gave bigger increases in yield than 100 to 400 pounds of dry fertilizer.

In a 31-acre wheat field near Sunbury, Ohio, the grower compared 15 acres where he applied 400 pounds of dry fertilizer with 15 acres on which he sprayed 2 gallons of 7–14–7 per acre. One acre between the fields received nothing. When we checked the field, we took the grain from a .02-acre plot as he harvested the field. We found that the dry fertilizer plot yielded 1.8 bushels more than the check plot, while the plot with the foliage spray yielded 6.9 bushels more. When he questioned our method of taking records, I told him that he should have 75 bushels more where he sprayed the foliage. When we got all his mill receipts together, he found that he had 67 bushels more from the 15 acres that he had sprayed, or 4.47 bushels per acre more than with dry fertilizer.

It has been my experience that when you finally get a grower to co-operate, he will apply the liquid on the poorest part of the field. In the above case, the half of the wheat field that was sprayed had several acres of Canadian thistles in it. The other half had no thistles.

Foliage sprays do no good where the calcium in the soil is low. We have a lot of soils in Ohio that will test neutral at some times during the year, but still are deficient in calcium. Many of these fields produce a low yield regardless of the amount of fertilizer that is applied. Fertilizer solutions will usually pay their way when used in these soils, but foliage sprays are wasted until we add more limestone to the land.

Tomatoes, peppers, and all members of the cucumber, melon, and squash family respond to foliage sprays. But if the soil has plenty of plant food to supply the needs of the plant, foliage sprays won't increase yields. We also have to pay attention to the stage of crop growth at which sprays are applied. Late fall sprays on wheat usually give good results because they tend to increase stooling and help the seedlings live through the winter. I have seen new seedlings of clover and alfalfa sprayed right after the grain was removed, making the difference between nothing and a good stand of clover. This may happen when seedlings are struggling to get a hold during hot, dry weather.

On a farm in central Wisconsin, I had exceptional results with seed treatments when only one quart of 10–20–10 fertilizer solution was applied to 2 bushels of seed oats. When the oats were harvested, the seed treatment increased the yield 24 per cent.

A number of years ago, one experiment station published results from an experiment on corn and soybeans in which they compared this one quart of solution applied to the seed with 400 pounds of dry fertilizer. They condemned the solution fertilizer because it did not produce more than the dry fertilizer. The use of the dry fertilizer was not condemned even though it did not increase the yield. One wonders sometimes how ridiculous some of our scientists can be.

We can't expect a batting average of 1.000 when we work with fertilizers. If we happen to get negative results in the first experiment, we should not condemn the experiment. Had we done this in the first dry fertilizer experiments that we ran, we would not be using dry fertilizers today, nor would we be using fertilizer solutions.

One hundred pounds of 10–20–10 solution contains 10 pounds of nitrogen, 20 pounds of phosphorus, and 10 pounds of potash.

This contains twice as much plant food as 100 pounds of 5–10–5 dry fertilizer.

Solutions cost $10 an acre to use. Because they are 100 per cent soluble and more effective than dry fertilizer, less is used per acre. Fifty pounds is all that is needed to grow 100 bushels of corn under average conditions. The cost per acre is approximately $10—this compares with $9 to $20 an acre for dry fertilizer.

Dry fertilizer that costs $90 a ton (12–12–12) would cost you $13 to $30 an acre. The cost of fertilizer per ton is deceptive. It is the program that cuts fertilizer costs.

Dry fertilizer people use figures that show that it takes 1.5 pounds of nitrogen, 0.8 pound of phosphorus and 1.2 pounds of potash to grow a bushel of corn. On this basis it would require 1,500 pounds of 10–10–10 to grow 100 bushels of corn—at a cost of $70. These figures are furnished me by the Ohio Agricultural Experiment Station. My research work on more than 100 farms in Ohio shows that 25 pounds of solution on the seed and 25 pounds applied to the foliage will produce 100 to 150 bushels of corn at a cost of $10 an acre for fertilizer. I can prove this by comparative field plots.

Foliage sprays, when applied according to my recommendations, will give profitable yield increases. This practice is based on results published by the Ohio Agricultural Experiment Station on the value of manganese sprays on soybeans. They found that applying manganese sulphate in a spray on the foliage not only corrected manganese deficiency in the soil but also increased yields. A complete fertilizer solution, among many things, has manganese in it and, therefore, will do the same thing. The practice of weed control by chemical sprays also shows that plants will absorb chemicals through their leaves.

I recommend that foliage sprays be applied when evaporation of moisture from the foliage is zero. The North Carolina Experiment Station, in a paper presented at the Cincinnati fertilizer meetings several years ago, showed that good results can be expected from foliage-applied sprays of fertilizer solutions if applied anytime after 3 P.M. and before 9 A.M. We have experimental results to show that when sprays are applied anytime when the humidity is near 100, the results are profitable.

I know that enough limestone applied to the soil will produce maximum yields. This may run into high tonnage on some soils. The Ohio Agricultural Experiment Station has for the past two years agreed to this practice, even on our natural limestone soils.

I found some twenty years ago that the application of adequate limestone releases phosphorus and potash to the plants. During the past year, the Ohio Agricultural Experiment Station has shown that available phosphorus may be increased by as much as 200 per cent. We have increased the available phosphorus by as much as 100 to 150 per cent on farms where our program has been in use for six to eight years and where no dry fertilizer has been applied during that time.

We have quite a few growers every year who get good results from our program but refuse to use it the next year. They won't tell you the reason for their decision. We know several reasons; there may be more.

Reason No. 1: "I talked with a man from the Experiment Station and he says your program is no good and it won't work." As far as authority is concerned, the grower may as well talk with his barber. Neither one knows because neither one has done enough research work to know whether it will do the job or not. They don't even have proof that they can get results with dry fertilizer. I spent thirty years in fertilizer research work and I know they have no proof.

Reason No. 2: "My neighbor makes fun of me because I use your program." All I say then is, if you want to run your farm in the red, that is your privilege. Our program is set up to make more profit. Remember how they laughed at the Wright brothers, at Henry Ford, at the kid who was fooling around with a homemade radio? My neighbors, when I was a kid, laughed at me because I tested seed corn for germination, because I read books on science, and later because I worked with fertilizer solutions. Perhaps I should have quit, but when people quit trying, progress stops. The people responsible for our progress were all laughed at while they were pioneering. When a man laughs at someone else's expense, he is showing his ignorance. We should laugh at humor, not at someone else's mistakes. I heard an army captain say one time, "Be careful how you treat the enlisted man. He may be your boss sometime." Anyone who laughs at your mistakes may want you to pull him out of a ditch sometime. We have several big customers who laughed at neighbors who used our program first. They don't want to be reminded of it now. Some of our best farmers have been ridiculed because they dared to try something new.

Reason No. 3: Some people who don't like to pay their bills jump from one company to the other to get credit. We have found that when you give a man extended credit, he becomes your enemy. I think many people have found this out. If you want to test a man's friendship, lend him a hundred dollars, and see how long his friendship lasts.

I have said before that no two people are alike. I am glad it is that way. What a monotonous life we would have if everyone reacted as you expected he would. Remember that the Roman Empire fell because people had no problems. Life was too easy. Adversity breeds greatness. Complacency leads to ruin.

So, we can expect everyone to have different reactions to anything new. My main concern is to prove to people that this program works. If they have plot comparisons on their own farms and get better results from our plot over another, it is about all we can do. If a grower doesn't believe what he sees, I don't know what more can be done. After all, if a man doesn't want to improve his standard of living, we must keep in mind that this is a free country; he can still do pretty much as he pleases as long as he doesn't break a law. We must also keep in mind that it is pretty difficult to absolutely prove many of the facts about growing crops. Things can be absolutely proved or disproved only if we know all the factors involved, but this, our mathematicians tell us, does not happen very often.

There is a misunderstanding about overliming. This applies

only to hydrated lime or Canadian wood ashes. As long as we use pulverized limestone, we don't have to worry about overliming. We have growers who have applied over 100 tons of limestone per acre and grow beautiful crops. We also hear about alkali fields, because they have a high pH. They usually need 5 to 10 tons of limestone to make them grow a crop.

But fertilizer solutions are no more than dry fertilizers the answer to our production problems. Our number one problem today lies in the inadequate use of liming materials, because without adequate liming, neither solution nor dry fertilizer can make us any money. With adequate lime (base saturation), as recommended by Growers Chemical Corporation, we can make money using both solution or dry fertilizers; but we can make it with less labor and less worry if we use fertilizer solutions.

CHAPTER 11

Nitrogen Depends on Sunshine to Be Useful for Increasing Crop Yields

IN THE EARLY history of our civilization we had sun worshipers. Life depends on sunshine. Whether the worshipers realized this is questionable. It is a happy coincidence that sunshine is free. Nitrogen is all around us as a gas and it is free, except that we have to change it slightly so that we can apply it to plants in a form they can get at. Some plants can make use of the nitrogen in the air. Some of this nitrogen also becomes fixed in the soil. Both nitrogen and sunshine are inexhaustible, for all practical purposes, but we must learn how to get the most good from both nitrogen and sunshine. If nitrogen is too abundant, it can cause us trouble. Nitrogen and sunshine are our source of proteins and amino acids in plants.

Leguminous crops like beans, clovers, and alfalfa, as a result of evolutionary processes, have joined company with certain bacteria which live in root structures so that indirectly, legumes use the atmospheric nitrogen for their growth. All non-leguminous crops must have nitrogen supplied to them either as a foliage spray, as nitrogenous salt, or as gases applied to the soil. In most cases soils can be treated so that sufficient nitrogen is available to grow a crop. Considerable nitrogen is applied to the soil with rainfall. The electricity of lightning changes the gas to salts, which in turn are carried into the soil by rainfall. This amounts to considerable quantities some seasons. Also, there are bacteria in the soil which seem to be able to change the gaseous nitrogen into proteins in their bodies. When they die, the proteins through oxidation become available to plants. Up to a certain volume of growth we have a system that will supply the world population for years to come, providing we don't expect too large a volume of growth from a certain area. Calcium is needed to make the chemical process work efficiently. We are assured, with a little co-operation on our part, of being able to feed people for many generations to come with what nitrogen nature fixes in our soils. All scientists have to do is find ways and means to make the chemical process in the air and soil work efficiently. I am sure that, since we have the means to apply sufficient calcium—taken from the vast stores of limestone all around us and ground finely enough—we can depend on having yields equal to more than our average yields without any further applications of fertilizer. This does not mean that fertilizers would not increase yields. It simply means that we are blessed with a permanent plant food supply that needs to be made available providing we don't interfere too much with crop growth. We have interfered in many cases by dumping a lot of fertilizer on land that didn't need it. So much nitrogen has been applied that many farm wells have been condemned because of the nitrate content. This nitrate begins to change to nitrite forms which are toxic to animals and humans.

My main concern here is the nitrogen and oxygen in the soil, because they are probably the main consideration in comparing yields from high and low organic soils. Up to the present it has seemed easier to get 175 bushels of corn from a mineral soil which is classed submarginal than to get that yield on a black prairie soil which is potentially 300-bushel-an-acre land. The type of growth on the mineral soil is more compact, less voluminous, and more fibrous, which, in the growth of the stalk, makes it sturdier. The color may be a rather grayish green. Corn on high organic matter has a weedy appearance. It is dark green, almost black-green; the leaves are larger and the stalks are taller. The stalks are weaker with far less fiber in them, the roots are not well developed, and the ears are apt to be smaller and unevenly filled. Because of this type of growth there isn't much that can be done. The plant absorbs too much nitrogen for it to utilize with the sunshine available. Until we know more about how to make a plant use a higher percentage of the sunshine that floods the leaves, we must do a better job of controlling the nitrogen supply. Crowding the plants may help to a certain point as long as the crowding does not cause sterile stalks. This type of growth requires more rainfall because of its greater succulence. Such plants do not have the fiber to strengthen them, and usually tip over soon after maturity. This is due to the fact that the plants are absorbing too much nitrogen for the sunshine that they receive. If the season is unusually cloudy, the plants are weaker, because the lack of the sunshine has the same effect as adding more nitrogen. Grape growers speak of bad and good vintage years. A lot of bright weather makes for good-quality wine because there is plenty of sugar in the grapes.

The amount of nitrogen needed by a crop depends on many factors. It must be used according to the sunshine received. Corn growing on the south slope of a steep hill can use more nitrogen to advantage than that growing on the north or shady side of the hill. Pineapple and sugar cane growers found this out many years ago. They apply nitrogen according to the sunshine they get.

The use of nitrogen by the plant is a complicated process and involves many changes. Few people who use nitrogen for growing crops have no more than a hazy idea why they use it. They seldom distinguish between nitrogen, phosphorus, potash, and complete fertilizer. Nitrogen, with the help of sunshine, by means of the green coloring matter, becomes a protein. The changes are as follows. Atmospheric nitrogen from the air is first changed to nitrous oxide by means of lightning and is washed into the soil as rain water.

Nitrogen and electric spark;
Nitrogen and legume bacteria;

Nitrogen and bacteria in the soil or ammonia added to the soil:
Eventually all this nitrogen is converted into nitric acid.

Nitric acid is neutralized by limestone and becomes—
Nitrate nitrogen in the form of calcium nitrate or ammonium nitrate, which—
Enters the plant root as a nitrate form.

In the roots nitrate is changed to ammonia by means of a plant enzyme;
The ammonia neutralizes an organic acid (which comes from sugar made in the leaf of the plant), which forms—
Amino acids—the building blocks for proteins. (Storage proteins are a source of energy for animals.)

Thus the nitrogen becomes part of the plant sap and functions as fuel does for an engine.

While this process is taking place, sugar is needed to supply energy and by-products to keep the nitrogen assimilation process going. At the same time the following process is going on in the leaves.

Carbon dioxide from the air, plus—
Water taken into the roots, plus—
Chlorophyl (green coloring) in the leaves and sunshine, build—
Sugars, starches and fiber.

Somewhere in the process some carbohydrates (sugar and starch) go—

1. To furnish stored energy
2. To make protein
3. To make fiber to give the plant strength
4. To form roots
5. To store in seeds, tubers or bulbs in the plant to produce a yield

No. 1 comes first. Then 2 is satisfied, as long as nitrogen holds out. Then 3 and 4 are taken care of, and if there is any left, it goes into building seeds and storage organs like tubers and bulbs. It is this surplus energy that, stored, makes our yield.

In other words, nitrogen is very important whether it comes

from what nature supplies or what man applies. If nature does not supply enough nitrogen to produce the maximum yield, then it is up to the grower to supply some, either as anhydrous ammonia, urea, or ammonium nitrate. If, however, there is enough nitrogen in the soil, which might well be the case in heavily manured or high organic matter soils that are well limed, then the extra nitrogen the grower puts on might well be harmful to his crops. It could do damage in the following ways—particularly in a season of normal rainfall. (If rainfall is so light that crops won't make much growth, it may not do much damage.)

1. Too much nitrogen could make a weak root system. It reduces proper feeding in the soil.
2. Too much nitrogen could make a weak stalk which would cause the grain to lodge or break over when maturing.
3. Too much nitrogen could use up so much starch that the pollen in flowers might not get enough. This would mean sterile flowers and reduce the kernels on an ear of corn, a head of oats or wheat, or form small potatoes.
4. Too much nitrogen could make certain crops bitter in flavor, so that animals would not eat pasture grass or hay. It will make fruit sour. It will make cucumbers and melons bitter. It will keep apples from getting red. It will make strawberries so soft that you can't ship them.
5. Too much nitrogen makes some crops mature slowly. Too much nitrogen makes corn shrink heavily in storage.

I made a preliminary check to estimate the yield on one corn field. It was a river bottom field, and by the size of the stalks and size of the ears, we knew the corn plants had access to more nitrogen than they needed. The ears were starting to dent. We counted the large ears on 100 stalks in various parts of the field and, from 90 out of 100 stalks having large ears, we estimated the yield and decided he should get at least 150 bushels. When the corn was harvested, it yielded 82 bushels. The grower said he couldn't figure out why there were so few big ears. I told him that he could have expected that, because there was so much water in the cobs that, when the corn ripened, the water was gradually driven off, causing considerably more than normal shrinkage. The sugar, starch, and fiber become watery, which may cause an ear to weigh a pound—but 50 per cent of that weight is water.

Just to show how this works, I helped check two fields of corn, both of which seemed to have exceptional possibilities for high yields. The first field checked out at 208 bushels of corn, with 26 per cent moisture in the kernels on September 1. We did not check the shelling percentage on either field. This soil needed considerable lime but had been manured heavily. Rainfall was good. On October 10, when the corn was harvested, the yield of No. 2 corn was 138 bushels. This was a high shrinkage. Another field which we checked at the same time had an estimated yield of 217 bushels but, when it was harvested, yielded 194 bushels No. 2 corn. Shrinkage was very light. The field was heavily limed but no manure or nitrogen had been applied. This corn was properly ripened, and would keep under most conditions of moisture.

Since nitrogen is so important in so many ways, it is important to explain what happens under varying growth conditions, because it makes so much difference in the quality of the ripened product. Unlimited nitrogen, water, high temperatures, and too little calcium can spell disaster to quality in many crops.

On one farm, 100 acres of melons tasted like green cucumbers when they were ready for market. When the night temperatures dropped below 60 degrees and the humidity dropped below 40 per cent, the melons developed a reasonably good flavor. The following explains what happened. Nitrogen taken in at first forms amino acids, which are soluble. They are associated with water, especially when calcium is low. A process of water removal takes place, and many amino acids merge to form a protein molecule and separated water. The water is not combined, and the protein is no longer soluble. This reaction is a ripening process, as chemically bound water molecules are released and the dry matter (starch) becomes more concentrated. Amino acids and sugar, as well as some starches, have water molecules diffused through the sap. When a kernel of corn is formed, it contains a very thin sap. As growth proceeds, amino acids and sugars are transported to the kernel, and the sap begins to thicken. As it becomes quite concentrated and begins to become milky, starch is beginning to form, and the sap becomes less soluble in water. Some of the sugar is condensed to starch, and amino acids, formed from nitrogen, begin to condense to proteins. A kernel of corn then begins to release water, and maturation has started. Less water is held in a chemical state. Such a kernel, if dried, will grow—but the seedling may be weak. The dough and dent stages are advanced ripening stages, as more proteins and starches are being formed and more water is released. The interesting thing to remember is that there must be a surplus of starch. If the plants keep on absorbing nitrogen freely, succulence in the plant is maintained and starch is slow to accumulate.

I was asked by a peach grower how to prolong the harvesting season of Alberta peaches, because he wanted to sell them at a roadside market. I suggested that he first apply considerable limestone to the soil. Then we set aside five trees in a block. The first block received 2 pounds of nitrogen per tree, the second 4 pounds, the third 6 pounds, and on up to 10 pounds per tree. The result was that the more nitrogen the trees received, the later the fruit matured. The last ones ripened six weeks late, and were rather bitter, because the nights were too cool to ripen them properly. You can get similar results with any crop. The presence of sufficient calcium tends to ripen the grain or fruit in shorter time. It tends to drive water out of the tissue. The type of proteins in hay have a lot to do with the curing of hay. Hay crushers came into use because the soil was low in calcium and too heavily fertilized. Hay grown on well-limed soil doesn't have to be crushed to make it cure properly, because it gives up its water readily. In other words, the dry matter that is built up—which is protein (insoluble), starch, fiber, and minute amounts of minerals—is the part that makes up our yields of high-quality crops. When dry matter is produced under conditions of high nitrogen, water, and high temperatures, it is made up of amino acids, sugar, and a watery starch which holds water by chemical bonds; it is slow to ripen because there is much water that has to be released.

When Darwin wrote down his observations of the rain forests of the tropics, he stated that he saw very few flowers, because the excess moisture, high nitrogen, and weak sunlight prevented plants from flowering and, therefore, no fruit was produced. They had no accumulation of condensed starches or carbohydrates. We have noticed this in wet seasons here in the United States. Weeds grow fast and succulent, just like the crops, and they produce very little seed. The flowers very often are incapable of setting seed or fruit. (Nitrogen might be a good weed killer if it were applied freely to a weed patch.)

When tomatoes make a very vigorous growth, they often will not set fruit. Ordinarily, when a tomato plant doesn't set fruit on the first flowers, the rate of vine growth is speeded up, which makes it even less likely to set fruit. In cases like that, fields that should have yielded 20 tons of fruit do not produce 2 tons of tomatoes.

The careless use of nitrogen has greatly reduced the world's food supply. More attention to and understanding of the place of nitrogen in our crop growth can make a big difference, and the use of just enough nitrogen could double our present food supply. We can't do anything about our sunshine, but we can learn to use nitrogen so that there is enough sunshine to go around. I am convinced that the reason I can grow 150 bushels of corn on "submarginal, worthless hills" is that the nitrogen is there in sufficient amounts to leave enough starch to produce that yield.

What is the relationship between phosphorus, potassium, calcium, magnesium, and many other elements, and nitrogen assimilation? They all serve a purpose. In the above discussion I assumed that we had an ideal situation—high enough temperatures, sufficient rainfall, and no obstacles in the soil that would prevent the plant from absorbing the nitrogen. The advantage of growing a plant in sand or gravel culture is that one can control the growth and can add or withhold any element. The reason one can't do this in soil is that there are obstacles which must be inactivated to keep the plant growing properly.

Growth is the result of two parallel chains of chemical reactions which support each other. One is the assimilation of mineral and organic ions, and the other is the assimilation of carbon dioxide in the manufacture of stored energy, which in the long run constitutes our yields.

Phosphorus and sulphur go through similar reactions associated with the nitrate ion. Phosphorus becomes a part of the proteins and nucleic acids, which are necessary to start the storage process of starch and which are also found in the nuclei of the cells. Without them we probably would not have growth. Proteins, with the help of phosphorus and sulphur (particularly in legumes and cruciferous plants), probably help to set up a buffer system in the plants which prevents rapid changes from taking place and serves as the base exchange complex in the plant, just as chemically active organic matter does in the soil.

Sulphur enters into the mustard oils and gives radishes, onions, turnips, and peppers their pungent quality. I assume that they enter into some proteins and such amino acids as histidine and lysine. It doesn't require much sulphur to keep a plant healthy.

Potassium, from potash, doesn't enter into any plant compounds. It has a minute quantity of radioactivity, which seems to be the active part of the potassium ion, and is supposed to serve as a catalyst in promoting certain chemical processes which have to do with the accumulation of sugar and starch. It also has something to do with control of the iron in the plant. When potassium becomes deficient in the plant, iron seems to become toxic and causes breakdown of the margined tissue of the leaves.

Magnesium is to the green chlorophyl what iron is to human blood. Without it, the plant turns yellow between the veins of the most mature leaves. The younger leaves show deficiency symptoms last. A deficiency of magnesium seems to interfere with the formation of the protection that plants have against sunburn. Of course, chlorophyl must be active, or the plant can't manufacture sugar, starch, oils, fats, and fiber.

I conducted an experiment in which I grew plants in sand culture in which I maintained certain nutrients at the threshold of deficiency for phosphorus, calcium, and magnesium, using lima beans as test plants. I harvested the seed and replanted them in pure sand. I grew them to the second true leaf. The results were amazing. Phosphorus deficiency showed an intense, dark-green color in the second generation. The seeds all seemed normal when planted.

Phosphorus deficiency: The seedlings were very uniform in shape and size. They grew 2 inches tall and stopped. They were a very dark green with a perfect growing tip typical of phosphorus deficiency.

Magnesium deficiency: These seeds germinated but had no growing tips. The cotyledons opened but no growing tip ever formed. This condition is referred to as "bald head." It is very common in some lots of beans grown from seed produced in the western states.

Calcium deficiency: An effect similar to magnesium deficiency. Some of the seedlings formed weak growing tips.

Boron deficiency: This is very similar to calcium deficiency.

Manganese deficiency: The symptoms affect the young leaves.

There are many symptoms not characteristic of any one deficiency which probably are the result of two deficiency symptoms. Plants growing in soil seldom show clear deficiency symptoms. Plants growing in a soil in which the calcium saturation of the base exchange complex is very low will exhibit a multitude of deficiency symptoms. Plants grown in a soil in which the roots are injured by a lack of air very often show nitrogen and phosphorus deficiency, probably because the two ions have to be assimilated in the young roots. Very often, therefore, a side-dressing of pulverized limestone will correct many different deficiencies.

I had occasion to advise a farmer on what looked like severe boron deficiency on small celery plants. I checked the soil, a sandy loam, and found it was very low in calcium, so I recommended a ton of limestone per acre applied broadcast to the plants and soil. All but the check plants recovered beautifully.

The check plants died. Since boron and calcium deficiencies are so much alike, it could have been calcium deficiency, but calcium usually is the last one of the elements to show deficiency.

In another case I had severe phosphorus deficiency on tomatoes that were just beginning to produce flowers. I had never been very successful in correcting phosphorus deficiency with superphosphate, so I dusted considerable limestone over the foliage and on the soil, and they started to grow freely in a week's time. Theoretically, perhaps, this is bad procedure; but if it does the job it simply means that the limestone releases the elements, increases the base exchange saturation, or corrects acidity. It all comes back to the idea that if we can saturate the base exchange sufficiently, the soil and the plant will begin to function normally and the plants will begin to grow. At the same time bacterial activity is increased and more nitrogen is made available. In black soils this could release so much nitrogen that plants could change from a hard type of growth to a lush, succulent growth, which might be a deterrent to maximum yields on that particular field. It might take several years before the nitrogen could be controlled sufficiently to get the maximum yield. This has happened on a number of Illinois soils where the limestone corrected the low calcium condition on high organic matter soils. In cases like this I would plant more seed corn to get more plants so that each plant would get less nitrogen. That would make it possible for the sunshine to be used more efficiently.

I have demonstrated many times that a big, lush, rapidly growing plant is not necessary to produce a big yield. I have grown ¾- to 1-pound ears—and sometimes 2-pound ears—on stalks not over 6 feet tall. Stalks alongside that were 8 to 10 feet tall did not have a good ear on them. My growers who have grown 175 to 195 bushels of corn an acre did so on stalks that were only 7 to 8 feet tall. When you can control the growth by controlling the nitrogen you can plant more seed on an acre. It takes a stalk to produce an ear of corn. Distributing what nitrogen you have among more plants makes it possible for the plant to make more efficient use of the sunlight it receives.

CHAPTER 12

Subsoiling and the Growth of Crops

MANY SOILS have lost their tilth through faulty management; the A_2 layer, just below the plowed A_1 layer, has become compacted. Even though this is a geologic formation, it has been aggravated by the soil being worked when there was too much moisture in it, partly because the calcium has become depleted in the base exchange complex and partly because of salt accumulation from fertilizer. During the formation of a sandy or sandy-loam soil in humid climates, and especially where it is derived from non-limestone rocks, plow soles are formed. These prevent root penetration and free vertical movement of water and salts. The clay becomes hydrated (a lack of calcium) and rain water begins to move it down out of the surface. Over the years this tends to accumulate between 4 and 12 inches deep, leaving the sand, silt, and coarser materials in the surface.

When man started to cultivate the soil, he found it plowed easiest at the 4-inch to 8-inch level, and every time he plowed at the same level the bottom of the plow acted much as does a trowel smoothing concrete. This tended to build up a plow sole which became more dense as the years went along. It could have been prevented had more limestone been applied.

I saw a 40-acre field in southern New Jersey which apparently had been plowed the same depth in the same direction for many years. The entire plowed layer had washed off during a hurricane that deposited 11 inches of water during a seven-day period. The water could not penetrate the A_3 horizon, so that when the surface soil became saturated with water it moved with the water, leaving a corrugated surface showing the location of every furrow bottom. See Figure 5.

FIGURE 5

A FINE SANDY LOAM SOIL WITH A SILTY CLAY SUBSOIL

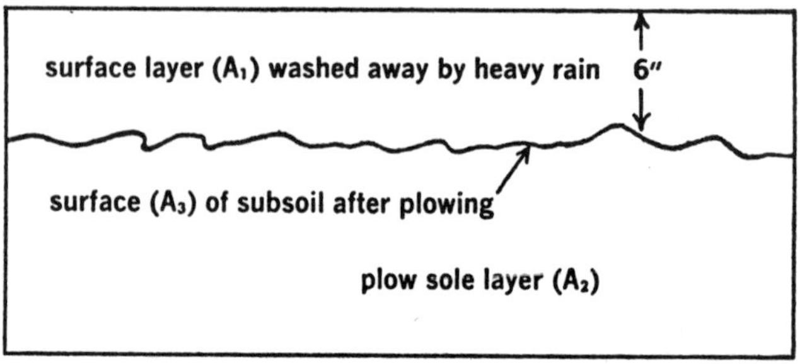

This soil was very compact. When dry it was almost as hard as a brick. When wet it was very sticky and smeary. The available calcium in this layer was less than 400 pounds per acre-foot. The soil under this plow sole was dust-dry right after the surface soil was washed away.

The low calcium saturation of the concentrated clay layer permitted potassium, sodium, and possibly some ammonium ions to take the place of calcium in the exchange complex. Since these ions have many more water molecules attached in chemical combination than do ions in the proper calcium saturation, it gave the clay a jellylike consistency which completely filled in the space between the larger particles. The near-colloidal solution readily moved out of and through the surface soil and gradually sealed the subsoil.

This plow sole condition could be corrected by applying liberal amounts of limestone and plowing it under along with deeper plowing. If the ground is too hard to plow deeper, it will have to be plowed early in the spring or broken up with a subsoiler. Plowing deep when too wet will puddle the clay and cause it to dry out in hard lumps. Several tons of limestone should be applied to the plowed ground. It will take a hard freeze or extremely dry weather to restore the structure, but it will eventually result in a better surface soil.

If limestone is not applied after plowing and turning up this plow sole, the ground won't grow very good plants, because the clay does not have sufficient saturation of calcium. The soil may act sterile. Some seeds won't germinate in such soil. I saw spinach growing on such land, where 10-foot sections of rows were completely devoid of seedlings. Limestone was broadcast over the tops of any seedlings that were up. In two weeks, seed germinated in the blank areas and the spinach made a good growth, even though the field was very uneven. Spinach makes a crop in eight to twelve weeks. There isn't much chance for late-germinated seedlings to catch up in a short growing season.

I was called in on a conference which was supposed to discuss overliming injury on corn in eastern Virginia. It was a case of one half of a field which had been limed with 1,600 pounds of hydrated lime eleven years before. The other half of the field was not limed. There had been a very wet spring. None of the corn, which was 6 to 8 inches tall, looked good, but the side that was limed eleven years before showed practically every deficiency in the book. A back furrow divided the two halves of the plot (one-half acre in all). From the appearance of the plants, I knew the roots were bad. I asked whether they had examined the roots. The man in charge said, "No." I am always amazed at how quickly people will jump to conclusions and at the aversion people have to digging around plants to examine the roots. Being a southern gentleman, he wouldn't get his fingers dirty. I dug up plants in both halves of the plot and laid them on a sheet of paper for comparison.

The good plants had good roots. The poor plants had no good roots. Even the seed was rotted. Also, the soil where the roots were bad had considerable red subsoil mixed with it. Where the plants were good, the soil was a brownish gray. I asked him why the soil varied so much in color. He said the side that had the lime eleven years before plowed so much easier that the furrow was deeper. I told him he had turned up sterile subsoil (low in calcium) which was beginning to kill off the seedlings. When I told him that if it were my field, I would apply 4 tons of limestone and run a subsoiler down 16 inches to correct it, he was so confused in his thinking that he didn't ask any more questions. I had a pretty good idea what he was thinking. I had been called lime-crazy before.

After my experience with plow soles in coastal plain soils, I moved to the Midwest and found myself involved in much heavier soils. I assumed I could forget plow soles. The few sandy soils which I found had plow soles as I had anticipated. But I was surprised to find a plow sole condition in the heavy soils. I found that the reason we had floods in Ohio, Indiana, and adjoining states was the dense condition of the soil, which prevented rainfall and snow water from penetrating the soil. I concluded that we needed this water for crop production. We had to find means of keeping it from running off. It had to be stored in the subsoil for future use by crops. The soil conservation people were working in the right direction but, instead of moving the water down, they were trying to slow it down in its flow, so it wouldn't erode. I found that plenty of limestone followed with a subsoiler which was pulled crosswise of the slopes reduced the runoff to a minimum. It held the water for future crop production.

Every time I visit a farm where crops are not growing well, I ask for a shovel or spade and have a hole dug. I also have a probe which gives me much more information. It is surprising how interested farmers become when they start to dig. Very few know what their soil looks like below the plowed layer. Since I have yet to find a problem soil which did not have some degree of compacting in the A_2 soil layer, or below the plowed layer, I assume it has held our yields down appreciably. In my mind it absolves farmers from any blame for low yields. They are guilty of having aggravated the condition, but even here I feel the blame should rest on the agricultural experiment station people for not giving the farmers better information. It looks to me as though we are in a rut where the blind are leading the blind.

I once complained to my brother-in-law about the way my

children were behaving. He said, "Training children is like teaching a dog tricks. You have to know more than the dog."

The information that I have depended on to correct soil problems was published long before I even went to college. Most of it was hidden in Russian scientific literature. If it had not been for my good friend Dr. Jacob Joffee, who translated much of the Russian literature for me, I probably would not have been aware of this subsoil problem. As it is, even though I feel very humble and inadequately informed on the subject, I feel that if I can demonstrate to a farmer that he can grow 150 bushels of corn on a submarginal hillside where the soil is low in organic matter, my information must have been reliable.

When I first contacted Dr. Joffee, we were working together on a tomato survey in which we maintained a close working arrangement with 132 farms for a period of three years. Our part of the job was to study soil conditions and fertilizer in relation to yields. We dug one or more holes 2 to 3 feet deep in each field so that we could study the profile. From the condition of the soil as shown by the profile, we estimated what the first yield would be, even though the plants had not shown flower buds. I jotted down the estimated yields as he gave them to me. After harvest we correlated actual yields ranging from 1.63 to 17 tons. For the three years our correlation was over 90 per cent correct. The soil was judged on the basis of appearance, feel, general moisture condition, odor, the compaction of the soil, along with the soil type. Although the relation of type of soil to yield was not too well correlated, we made chemical soil tests on all the farms. These tests did not enter into our estimation. We found no correlation between fertilizer applied and yields. We did find a very definite correlation between the dollars a man spent for limestone and yields. Since the limestone affected the soil, it was easy to see why Dr. Joffee's estimates were so close.

From 1946 to 1949, while I was at the Virginia Vegetable Research Station, I co-operated in a state-wide test of some 72 varieties of field corn. Every year a few new ones were added and an equal number were discarded.

I had limed and subsoiled the field the year before. The subsoiler furrows were 3 feet apart and 21 inches deep. The corn rows crossed the subsoil furrows. When these varieties were harvested, I found all varieties had ears ranging from 3½ to 11 inches long. By observation, I soon found the stalks with long ears lined up in rows crossing the subsoil furrows. Mr. Cummings of the Engineering Department at the U.S.D.A. at Beltsville made me a steel probe out of quarter-inch rod pointed at one end. On the other end just below the handle he had attached a pressure gauge that registered up to 200 pounds.

I marked 1,000 stalks (the seed was all planted by hand—8 inches apart in the row) in a block and made a measurement alongside each stalk. I measured the length of each ear when it was dry enough to be picked. Then I pushed the probe down 12 inches alongside the stalk and jotted down the figure showing the pressure on graph paper to see the correlation between length of ear and pounds pressure required to push the probe into the subsurface soil. The inverse correlation between ear length and pounds of pressure needed to force a probe into the soil was 91.2 per cent. See Figure 6.

FIGURE 6

EFFECT OF SOIL COMPACTION ON LENGTH OF EAR

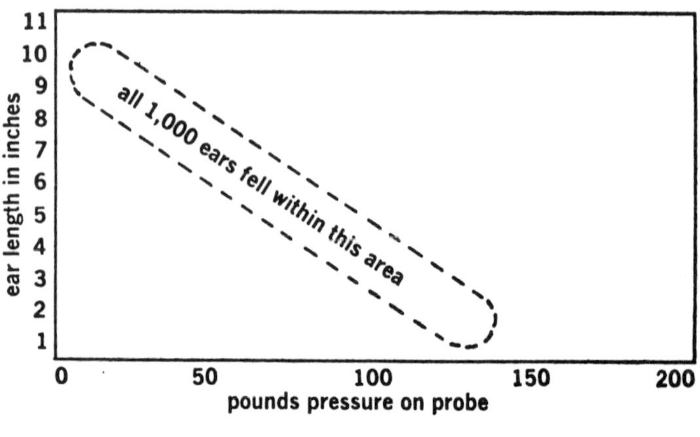

There seems to be an indication that the hard, dense subsoil affects yields. I dug the soil away from one side of plants with 4- and 10-inch ears. The short-ear plants had shallow roots mostly in the plowed layer. The long-ear stalks had their roots deep in the subsoil, where the cracks made by the subsoiler permitted them to get through the dense soil or plow sole. The roots not only got more water deeper in the soil, but there was a chance that the roots had access to much more air. Of course, it does give roots access to more soil to feed in. I have an idea that the better aeration of the plant roots probably had more to do with the larger ears than the additional nutrients.

There has been much scientific discussion of the value of subsoiling. I have resorted to subsoiling on many occasions and I have never seen any damage from the practice. I have had some cases where it did no good the first year.

I ran an experiment on muck soils in the Great Meadows area of New Jersey in co-operation with Mr. Charles Nissley. We subsoiled part of the beds, and in other rows we dropped limestone from a hopper on top of the furrow. This limestone partially flowed down into the furrow and part of it stayed on top. Celery plants were set on the land between two drainage ditches 100 feet apart. We got absolutely no response the first year and we gave up the experiment.

A year later, after the second harvest, a farm equipment dealer in the area called me and wanted to know if we were recommending subsoilers to the Great Meadows celery growers. He said he had orders for fifteen. When I checked into it, I found that the plots that were subsoiled two years before had twice as much celery as the check plot, which was not subsoiled.

I have read reports of negative results from subsoiling. But when I checked further into the manner in which the soil was subsoiled, I found the reason. These people wanted to improve drainage by subsoiling wet ground. It does no good to subsoil ground when it is wet. The blade cuts through, but then the soil runs back together. The soil also puddles when a tractor is run over it. It not only puddles the surface where the tractor wheels compact the soil, but it puddles deep, where the blade cuts through. The time to do the job is when the soil is dry and hard. Then the subsoil fractures the plow sole, causing it to break 3 feet or more on either side of the subsoiler.

The purpose of subsoiling may be for drainage. A mole is drawn in back of the subsoiler, leaving a round hole through which the water can flow to an open ditch. This is done on flat ground. It must be done when the soil is wet enough to be molded into a round opening which will hold its shape. If it is too wet, the soil collapses. I am not too much concerned with this type of subsoiling.

My purpose in subsoiling is to speed up the penetration of limestone into the soil. This hastens the corrective action of limestone on the plow sole by giving the water a chance to carry the limestone down to the plow sole where it can correct the unsaturated condition of the clay to make it possible for roots to penetrate. This use is for muck and heavy clay soils. When sufficient limestone has impregnated the plow sole, water moves down faster and less runoff occurs.

I don't consider subsoiling a yearly operation. It is a means of speeding up the penetration of limestone in the subsoil. It encourages roots to penetrate deeper, and, as a result, the roots have more soil to feed in. It also encourages chemical reactions to hasten the release of more nutrients for crops. After lime has sufficiently saturated the base exchange complex, subsoiling is no longer needed. Occasional applications of limestone will maintain an open subsoil. Subsoiling in low calcium soils probably will do very little good. Applying limestone along with subsoiling is a happy combination.

Another reason for subsoiling is to store water on high ground, for future use by growing crops. In addition to storage of water, the subsoiling keeps water from causing floods. Of course, this can do good only if there is a joint co-operative effort, so that all land is so treated. This would be a big project, but if it works on a small scale, it will work on a wide area, except where trees cover the hills and slopes. In areas along the coastal plain where the soil is only a few feet above sea level, there may be heavy clay soils. Since tile does not drain these soils readily, large ditches have been dug to carry off the surface water. The soil is a mixture of clay, silt, and some fine sand, and is very deep.

A grower whom I worked with on a coastal plain farm told me he could buy 160 acres of this heavy soil very reasonably. It had not grown a good crop in forty years, as far as anyone knew. He asked me whether I thought it could be made productive. I told him that as far as I was concerned, every soil could be made productive and that I would be glad to work with him on it. The farm was located between several highways, as shown in Figure 7.

FIGURE 7

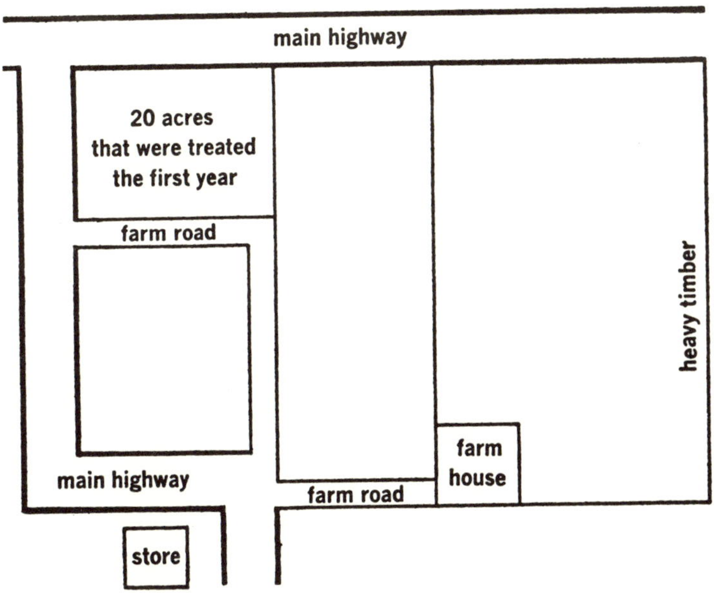

The east half of the east 40 acres was taken as a starter. In midsummer the east 20 acres was picked for the demonstration. The soil needed 10 tons or more limestone per acre. The new owner spread 6 tons limestone, plowed the ground as deep as he could, and smoothed the surface. We built a hopper on the rear beam of a mounted subsoiler, as shown in Figure 8.

FIGURE 8

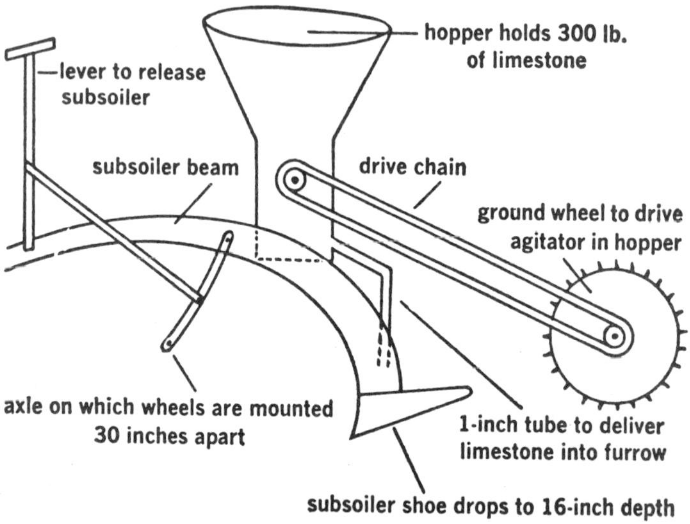

We applied 2 tons of limestone with this equipment and another 9 tons by subsoiling the ground crosswise as deep as the machine would dig. That made a total of 11 tons of limestone. Of course, we had the usual number of scoffers. The benefit derived from this treatment was little short of miraculous. Corn was grown successfully in this field the following year. Farmers could not believe that the soil could be improved in such a short time. The limestone mellowed the soil and the subsoiling made it possible for rain water to penetrate the soil with very little runoff. The subsoiling speeded up the correction brought about by the limestone.

In another case in Ohio a grower had a clay knob in a field which did not produce even a fair crop. It did have a lot of clay in it. When I tested the soil I found that the field needed 6 tons of limestone while the clay knob needed twice as much, even though the pH was near neutral. The limestone was plowed under. The next year the knob yielded 154 bushels; the next year, 137 bushels; and the third year, 143 bushels. This was not subsoiled, but the limestone accomplished the job of helping the roots to penetrate the subsoil. It is possible that, had this knob been subsoiled when the limestone was applied, the yields might have been even larger.

A spinach grower who usually grew 200 acres of spinach and produced approximately 300 bushels an acre, wanted to know why his spinach was frowned upon by commission men. I told him it was poor quality. He said it was as good as anyone else's in the area. I told him all spinach from his area was frowned upon. When I told him he should be growing 1,000 bushels of high-quality spinach an acre, he looked at me through half-closed eyes and said, "Tell me how to do it."

Thanks to my big mouth I was on a spot, but I was pretty sure I knew what the trouble was. He had a sandy loam soil. He had a bad plow sole and he was using 1,500 pounds of commercial fertilizer for each crop. He grew two crops a year. Dr. Danielson and I went to work. We found the soil was neutral. He had very little calcium in the base exchange complex.

We started applying limestone, and in three years had applied 7 tons per acre. We reduced the fertilizer application from 1,500 to 500 pounds per acre. The average yield of spinach the third year was 1,184 bushels an acre, and his commission man told him it was the best quality of spinach he had ever had the pleasure of handling. Figure 9 was constructed from data and observations made from this spinach field. (A *profile* consists of a slice of soil from the surface to a depth of three or more feet, showing variations at different levels.) Subsoiling was not used on this field, but it might have speeded up the correction had it been used when the limestone was first applied.

FIGURE 9

PROFILES OF A SPINACH FIELD

Left, before liming; right, after liming

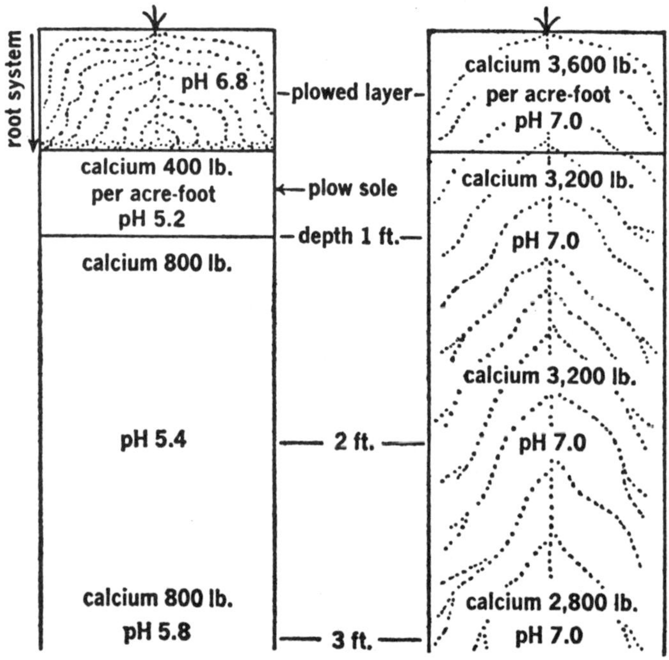

The importance of subsoiling, when done at the correct time, should not be underestimated. In 1960 some corn turned yellow when it came up. When it was 12 to 15 inches tall, we ran a subsoiler between planter rows in order to introduce some air into the soil. Heavy spring rains had compacted the soil and had sealed the surface. This soil needed close to 12 tons of limestone to correct the calcium deficiency. For this reason we decided that the yellow foliage was due not directly to nitrogen deficiency but to the inability of the plant to use nitrogen.

Two weeks after we subsoiled between the corn rows, the

plants had definitely changed their color to a deep green, while the plants left as a check were still yellow. When we harvested the corn, we had 53 bushels more corn than where we did not subsoil. The yield was 103 bushels, compared with 50.6 bushels from the non-subsoiled plot.

Thus, to summarize the problem of subsoiling:

1. It is purely a mechanical operation to help drainage, help water penetration, and help air exchange. Along with this we can expect a certain amount of surface material—organic or other—to be carried into the subsoil.
2. To get the most benefit from subsoiling, the job should be done when the subsurface reaches of the soil are dry and hard, so the ground will be cracked in many directions.
3. Pulling a subsoiler through wet ground probably is a waste of time, since it can result in a puddled condition. Under these conditions, air is at a premium, and any organic matter dropped into the subsoil probably will not decay, since the oxygen supply is limited. On this basis, I expect that the suggested practice of blowing shredded dry organic materials into the opening made with a subsoiler would be of little value in increasing crop yields.

The type of profile in Figure 10 is common in soils of limestone origin. The plow sole, if present, is not well formed and often is nothing more than a dense area in the A_3 horizon. This is even more common in the sandy loam soils. To adequately saturate these soils with calcium requires 2 to 10 tons of limestone. These soils have a comparatively high pH even though the available calcium reading may not be more than 800 pounds per acre.

FIGURE 10

PROFILES OF LIMED AND UNLIMED SANDY LOAM SOIL

Soil Origin–Basic

Left, soil with low calcium saturation; right, soil adequately supplied with calcium

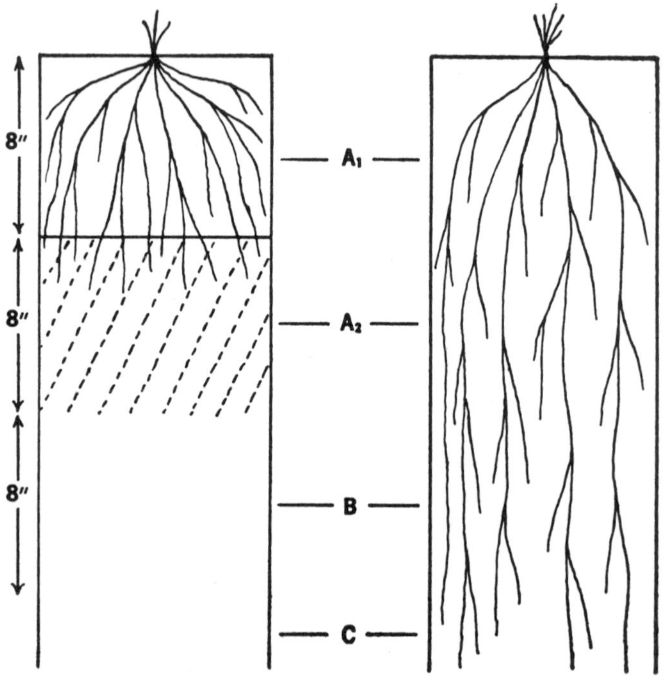

	Left		Right
A_1	pH 4.7–7.1 Calcium 800–1,600 lb.	A_1	pH 6.2–7.2 Calcium 2,800–6,000 lb.
A_2	pH 5.4–7.2 Calcium 800 lb.	A_2	pH 6.8–7.0 Calcium 2,800–6,000 lb.
B	pH 6.4–6.8 Calcium 400 lb.	B	pH 6.8–7.0 Calcium 2,800–6,000 lb.
C	Similar to B	C	Same as B

FIGURE 11

PROFILES OF LIMED AND UNLIMED SOIL

Soil Origin–Acid Rocks

Left, soil with low calcium saturation; right, soil adequately supplied with calcium

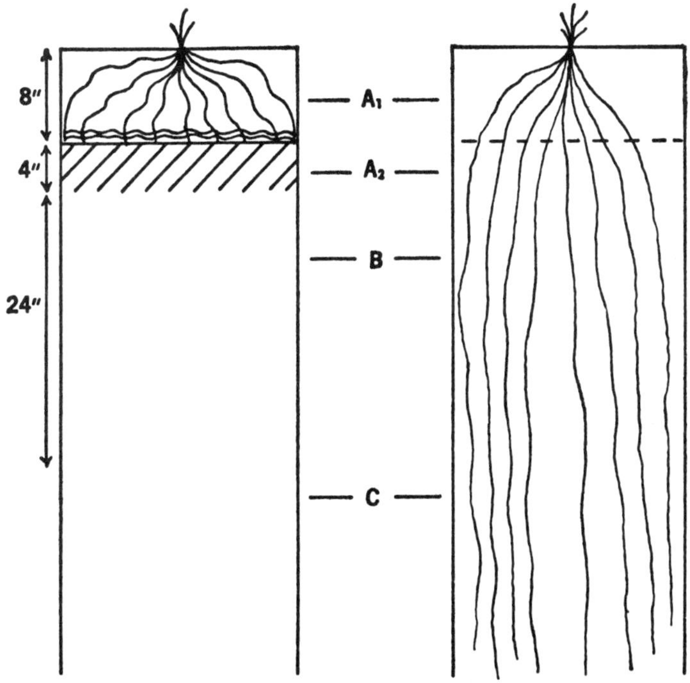

Left:

A_1 pH 4.7–7.1
Calcium 400–1,600 lb.

A_2 pH 4.7–5.4
Calcium 400 lb.

B pH 5.2–6.8
Calcium
Sandstone 800 lb.
Limestone 2,800 lb.

C Similar to B

Right:

A_1 pH 6.7–7.2
Calcium 2,800–6,000 lb.

A_2 pH 6.8–7.0
Calcium 2,800–6,000 lb.

B pH 6.8–7.0
Calcium 2,800–6,000 lb.

C Same as B

The type of profile in Figure 11, with considerable variation in structure and fineness of material, represents the condition found in the majority of our farm lands. This may be in the sandy types where the plow sole is most noticeable, common in coastal plain soils. It is less severe in the silt loams and only slight in the clay loams. To adequately saturate this type of condition to encourage roots to penetrate the full depth (5 feet) may require from 5 to 30 tons of calcium limestone, depending

FIGURE 12

PROFILES OF LIMED AND UNLIMED MUCK SOIL

Left, soil with low calcium saturation; right, soil adequately supplied with calcium

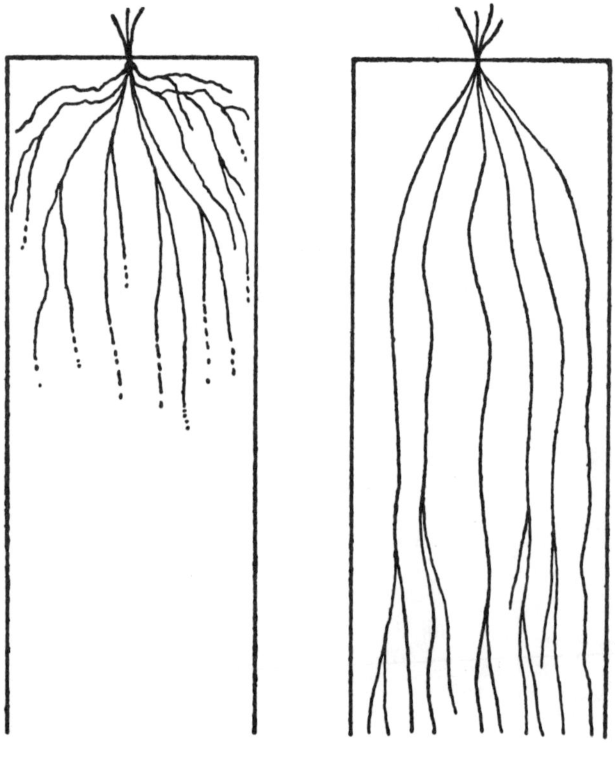

on the percentage of colloidal clay and amount of organic matter present.

The profile in a muck soil (Figure 12) varies tremendously. Actually a muck soil is nothing more than 6 inches to 6 feet of loose muck soil on a tight sand or clay soil. A muck soil may require 40 tons of limestone per acre foot, while the sand underneath may only need 2 tons and a clay underneath may need 10 tons.

A Weisenböden, often called Jackwax by farmers, is a dense, fine, compacted black soil which has a poorly defined profile and requires 5 to 30 tons of limestone per acre-foot. When adequately limed, these soils will become mellow and produce high yields.

CHAPTER 13

Soil and Plant Tests May Be Useful in Increasing Yields

EVERY SOIL and plant research worker has the ambition to have enough data and enough observations to perfect a soil test and a plant foliage test able to recommend the exact amount of plant food materials to grow the maximum yield. This is a big order. From my thirty years of experience and consideration of soil and plant tests I am ready to approach the problem with the utmost humility, knowing that I have an insurmountable obstacle before me. I have used soil tests for many years, and plant tests at spasmodic intervals. As a result of experience in using these tests on field crops, I have gradually soft-pedaled many of them, because I could see no correlation between my tests and field results.

The purpose of the soil test is to find out why a given soil does not respond to the application of those nutrients which the soil test shows are needed. If we consider the complexity of the media we are working with, we will see that there is a reason why the soil test is not always reliable.

A pure sand culture, if properly supplied with nutrients necessary to a good crop, will, when tested with a quick soil test, show a good test for calcium, manganese, nitrate and ammonia nitrogen, phosphorus, potassium, and various trace elements. If, unbeknown to the tester, we leave out one of these, he can find which one it is and, if he applies it, the plant will grow normally. It is a simple matter to correlate the results of the test with the growth response. In other words, the tests are accurate. The technique is satisfactory. We do seem to have the available testing equipment. And when a person asks me whether a given tester on the market is good, I must say it is a good tester. But there are many conditions where we find the tester useless because the results do not correlate with the resulting crop response. What, then, is the problem?

We have had many scientists studying soil testers, trying to adapt them to certain soil types, but so far I have been very much disappointed. I feel that the wrong people are working in the wrong places. The scientist who should be working on the perfection of the soil tester is the one who knows the chemical and physical changes that take place in a soil. A man with little knowledge of soil chemistry probably never will contribute much to the perfection of a soil tester nor even to the perfection of a plant foliar tester, if such a thing is possible.

We must assume certain conditions. We have tremendous changes taking place and different conditions prevailing in the many soil types, yet we are trying to adapt a soil tester to work for all of them.

Every soil has rather large quantities of minerals, and some have increasing amounts of organic matter. As such, they may have little value to our growing crop because the nutrients are not in a form capable of being taken into the plant. A sandy soil may be made up largely of quartz. To this quartz may be added more minerals which eventually may supply some or many of the necessary nutrients, in varying quantities. So far these nutrients are not in water-soluble form and may not be extracted from the sand with distilled water. They may be broken down later and may eventually be made available. Until we begin to reduce or oxidize some of the iron and aluminum minerals to a chemically active form we have a more or less inert material, as far as the plant is concerned. To be available to the plant, they must be extractable from the sand by distilled water.

The strongest means the plant has to make nutrients available from these minerals is carbon dioxide, which becomes carbonic acid, a weak agent but effective if time is not a factor. But usually the process by itself is not fast enough to make the bound nutrients available so that the plant can make a satisfactory growth.

A good soil test consists of an extracting solution strong enough to take out or release the available nutrients used by the plant in a given soil. The tests are then made on the filtered solution. The big problem is to find an extractant that will remove what is available to the plant and not remove that portion which is held too firmly for the plant to extract.

In the sandy soil mentioned above, distilled water would be a good extractant, and in most soils distilled water will remove some of the nutrients used by the plant. So far, we have been dealing in simple chemistry, and a knowledge of elementary chemistry might suffice. But when we consider the other 99 per cent of the soils in which our crops grow, we deal with colloidal and organic chemistry reactions which complicate our problem very much.

If we mix sand, certain necessary minerals, and a small amount of organic matter, we gradually form, with the help of carbonic acid from the roots of plants, a combination of quartz, iron oxide, and aluminum oxide which, because they are chemically active, begin to act as a colloid with negative charges which are ready to combine with positive ions. Then, if we add some limestone, a base exchange system forms and begins to control the mechanism of exchange between the charges on the active colloidal system in the base exchange system and the charges on the roots of plants, which get their charges from the protein system in the protoplasm of the root cell. Thus, we have established a strong base exchange complex in a soil which may be largely chemically active colloidally fine clay; or we can have protein-clay combinations, or a colloidal protein system in the root cells which also possesses base exchange properties which are probably weaker than those charges in the base exchange complex.

If we set up a field experiment on a silty clay soil which needs 8 tons of limestone, applying 2, 4, 6, and 8 tons of limestone, the yield of corn will be similar to curve A in Figure 13.

FIGURE 13

CORN YIELD FROM (A) LIMED SILTY CLAY SOIL AND (B) LIMED MUCK SOIL

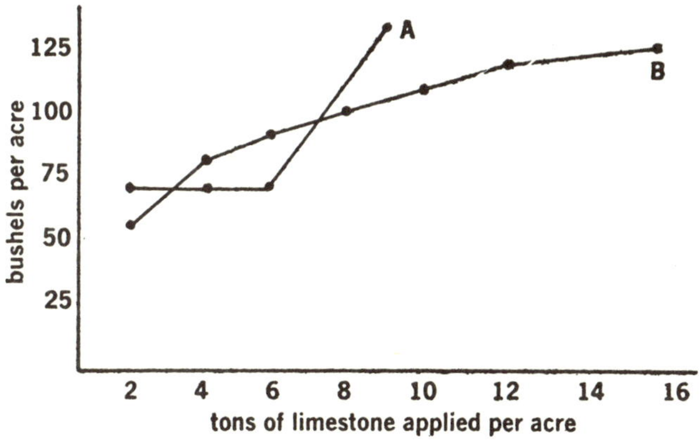

If we set up the same experiment in an organic or muck soil, the curve will follow the pattern represented by B. This is so because calcium is not released as readily from a mineral base exchange complex until 85 to 87 per cent of its capacity is reached. In an organic soil we get some response with each 2-ton increase, because the calcium, being more lightly held, does find its way to the root colloid.

Now, if we consider our soil test, what are we going to conclude? We make up an extracting solution that contains sodium ions. The strength of the extracting solution determines how many of the calcium ions are released. We want to approximate what the plant can take out. We would probably need a stronger reagent for the clay complex than for the organic or protein complex. When I say protein, I am including those organic compounds in the organic matter, other than proteins, which may have negative charges.

So we mix our extracting solution, 50 grams with 20 grams of soil (10 grams of a muck soil, because an acre-foot is one-half as heavy), and permit it to stand long enough for the sodium in our extracting solution to take the place of the calcium, magnesium, potassium, manganese, and other ions on the exchange complex. I have taken two extreme cases. We have all possible combinations, which makes our problem even more complex.

We haven't said anything about phosphorus. Phosphorus, in its various combinations with the heavy metals as well as with the alkaline earths, must be considered in terms of the pH of the soil. Under rather acid conditions, phosphorus probably is tied up with the heavy metals. As the acidity decreases, more and more phosphorus may be associated with calcium, magnesium, manganese, potassium, sodium, and even ammonium ions. It has been shown that as more limestone is applied to the soil, more phosphorus becomes available. Under these conditions, plants apparently can absorb more phosphorus or do it more readily and seemingly make a better growth with much less phosphorus than at a higher degree of acidity.

We might also assume that the phosphorus ion, because of its dual quality, could take part in base exchange reactions with the linkage through a bivalent cation.

I mention all these things because we are trying to perfect a test that will give results in the face of a million variables. Without seeming to be a defeatist, I do want to leave the impression that we may or may not have a useful tool. I have said time and time again that anyone can make soil tests and get a set of figures. But what are we going to do with the figures? There are people who use the soil test to good advantage. They know their limitations and the conditions under which they work and, by field tests, they have found a good tool.

My own experience with soil tests has led me to put a lot of confidence in a calcium and phosphorus test. With a definite philosophy, I have worked out—with reference to the importance of the calcium ion in maintaining good tilth, drainage, and aeration in the soil—its bearing on efficient utilization of other plant nutrients as shown by increased yields. I have placed most of my dependence on the calcium test. Next I would mention the phosphorus test. Beyond that I have little confidence in soil tests. I run other tests in my laboratory because it is easier to run the tests than to explain to people why I do not run them.

The potash test has been disappointing. In dry weather I get a high reading, while a sample taken in wet weather will show little or no potash. Fertilizer salesmen should pay attention. I look for other things. I always look for plow soles. I use a pointed, three-eighth-inch rod. With a well-established plow sole, the calcium test is the only test that tells me anything. I smell the soil, especially when it is wet. A foul odor means trouble, regardless of what the soil test shows. It means poor aeration and faulty drainage. I think we need soil tests for toxic materials. Nickel, chromium, and possibly other metals may cause troubles. Laboratories should be equipped with facilities to run some of these tests.

Soil tests are accurate enough when the test is made, but when one realizes that the soil is in a state of continual activity and probably changing by the hour, one wonders how to make the best use of the test. What possible chance do we have of getting a response to potash if we find it is low in the soil and we apply some? Certainly it is not 100 per cent. It may not be 25 per cent. We may not get a response at all because if the soil sample was taken after a heavy rain, most of the available potassium may have been carried below the zonal level that was tested. Perhaps a plow sole prevents roots from absorbing the potassium. If the soil became low in moisture before the potash was applied, the potash could be moving back toward the surface and, if we tested for it again, we might find an abundance.

We can take samples of soil out of one hundred different locations on different farms and we might find them all deficient (if anyone can tell us where the threshold for potassium deficiency should be). If we apply 500 pounds of potash per acre on each one, we might not get a response on any of them. We could get a response on 75 per cent of them. I doubt very much that we would ever get it on 100 per cent of them. If we get it over 50 per cent of the time, I would consider the soil in that area generally low in potash.

We find soils in certain areas generally low in some one nutrient, due to the fact that certain minerals which supply the particular element are nonexistent. I have worked in areas where soils were particularly high in magnesium, potash, sodium, and calcium. There probably will be no response to these particular elements under such conditions, should we find them low.

Rolling land varies from steep to gentle slopes. Such soils may vary from sandy loams to clays, but are usually comparatively low in organic matter, because they are well aerated and do not build up much organic matter, except where temperatures are low enough to slow down oxidation and the activity of microorganisms. In the bottoms of the valleys or depressions we have more organic matter.

The quantity of organic matter at any one time is the balance between the organic matter that is oxidized each year and what is deposited by root growth, what little roughage is turned under with the plow, and what may accumulate on the surface. Therefore, as we approach the equator, we may have a loamy sand, but when we approach the northern boundary of the United States, that same soil type would be classed as sandy loam with 1 to 2 per cent more organic matter, and if we should add to that above-normal moisture conditions, we could build the organic matter to 3 to 6 per cent.

Rolling lands may be hilly because of glacial action, in which case we find gravel and field stones in abundance, or because of erosion of mountainous areas, where we find fairly coarse soils on the slopes and fine soils in the valleys as a result of water covering the coarser materials with fine sand, silt, and clay. Lake or sea bottom lands usually have an abundance of clay in the central states. Our desert soils are the result of sea deposits which have dried up because a change in air currents caused rainfall to move to other areas. Some may be reclaimed with irrigation.

From the standpoint of the world food supply, these soils are important, but each one presents a different problem. A soil test may be of some help in central Illinois or Iowa, but without adaptation, that same test may be useless in southern Alabama and quite inadequate in New York and Pennsylvania, or even New Jersey. And yet scientists in the East argue that they have a better test than the scientists in the central states. They may even go so far as to say that the tests used by the people in the central states are no good. We have argued for many years about the adequacy of the soil test, and we may argue many more years and still not come to an agreement. There is no basis for agreement. At this moment, our soil test is not too reliable. It probably is accurate enough in itself; but how can we interpret results in terms of the many different soil types or varieties we may encounter in a given area?

There are keen observers among our scientists who depend as much on their senses as on actual soil tests to help them solve crop production problems. Soil tests are of recent origin; but people have grown crops for centuries and centuries.

Corn, one of the staple crops on which Americans have depended for their food supply, dates back more than two thousand years. I don't suppose there is any definite record as to where or by whom corn was first grown. There seems to be good evidence that it was grown on the fertile soils of Central America and Mexico by the learned Aztec and Mayan people before history was recorded. It is interesting to note that "big corn yields" were harvested by the Mayan Indians from those fertile soils, which were underlaid with limestone.

Five thousand years ago it was known that good corn could be grown better on the soils whose origin was in limestone rock. Some of our farmers today don't appreciate that fact. Instead of taking some lessons from our early American forebears and advancing their high level of knowledge, gained from a thousand years of experience, we ignored that and started from scratch. Actually, we have gone backwards to the point at which average yields are now ridiculously low. It is not so long ago that one heard the statement "Corn doesn't need limestone. Only legumes need limestone." Apparently we were thinking about growing corn without soil. It may be true that corn is more tolerant of toxic materials in the soil than some other crops, but it took us too long to realize that if we planted corn on heavily limed soil, we grew much better corn than we did on so-called acid soil. It was a case of the agronomists, with their limited knowledge of soil and plant chemistry, getting the jump on the soil chemist, who had more fundamental knowledge at his disposal. Our whole research program on crop production has had the cart before the horse.

We can say the same things about most of our food crops. Somebody classified all plants according to their lime requirements and our gullible crops research men swallowed the bait without batting an eye. Had we assumed that these lists were based on tolerance to toxic conditions in the soil and, had we considered the calcium needs of the crop and soil, we would have made much more progress. As it is, we still don't realize that the chemical and physical condition of the soil must be given first consideration in any program of crop production. We neglected to keep our soils supplied with calcium and now we suddenly realize that what we have taken out of the soil we must replace. This means we must apply large quantities of limestone. As a result, we are in a muddle, arguing about things about which few of us have sufficient knowledge.

If we were alert, observing, and open-minded, we could learn much through experience. A great teacher, Agassiz, once said, "Study nature, not books." Don't shackle your mind by what someone has said unless you are wise enough to know wisdom when you see it. I once met my freshman chemistry teacher on the street as we were going to his lecture. He told me virtually the same thing: "This is a wonderful university and we have fine libraries. Take advantage of them. Don't depend on just what you hear in the classroom."

A soil is an active chemical compound which must be in balance with the growing plant. Any chemically active material sets up electrical charges. The more clay and organic matter present, the more charges a given square foot will contain. A growing plant likewise has electrical charges in its protoplasm. In other words, if the soil is neutral, there is little interference with free ions moving from the soil solution to the root where charges are waiting to be satisfied. The root can also repel the entrance of ions into its cells, if it has its charges neutralized.

Any living thing in the soil which depends on osmosis for its nutrition undoubtedly is affected by these same laws. The complexity of all the processes going on in the soil and their effects on plants is staggering. How can we possibly grow a good crop on a soil if we know almost nothing about it? And yet our crop research has continued for seventy-five or more years with very little understanding of what is happening around the roots. And soil tests are supposed to rehabilitate the agricultural industry.

The term "fertility" as used now is too confining and has led us down a narrow alley. It has been a boon to the fertilizer companies. The word fertility, to me, means the correct balance between all the forces contributing to producing a maximum yield under a given set of climatic conditions.

If, for instance, I can raise the yield of corn from 50 to 103 bushels by squirting oxygen gas around the roots of corn, as I did on the Olena Farm in 1960 when corn became very yellow early in its growth, I feel that I have provided the plant with something it needed. In other words, oxygen had a beneficial effect and should be considered a contributor to the fertility level, just as we consider nitrogen and phosphorus. We need an oxygen tester, for if oxygen contributes to yield increases it becomes part of the fertility picture.

The big question is, Do we understand our soils well enough to know why oxygen increased the yield? Roots need oxygen. True! But why didn't the soil provide it? It is one of the things that contribute to high yields on some soils and not on others—while all of them may have equal amounts of plant food materials. Lime promotes high yields. Why? We have very few of the answers. How have we changed the soil? Is the drainage of water from that soil more rapid? Have we improved aeration? Have we detoxified something? We certainly have changed the chemical condition, because we added calcium and magnesium. Have we changed the structure? Have we released plant food? There are hundreds of things that could have been changed. How much do we know about the clay and organic matter? I will agree that it is easier to grow a good plant on sand than on clay soil, not because sand is better, but because we know more about pure sand than we do about clay soil. Sand lends itself to more misuse and bad cultural treatment. It has fewer things that can go wrong. It warms up faster. Water penetrates faster and carries oxygen into the soil with it. There are fewer electrical charges to be neutralized.

When a youngster, I made mud balls and mud pies. Certain things happened which I still remember and which tell me much about a soil. The kind of soil determines what kind of mud balls you can make. Perhaps it would help if many of our farmers were to add limestone to some of their soils, make mud balls, and see what it does to the soil. This reminds me of a county agricultural agent, a good friend of mine, who always carried a pack of blue litmus paper in his pocket. If we were in a field, he would pick up a handful of soil and make a firm ball of it. Carefully he would place the ball in his pocket and, after ten minutes or more, break it open, lay a piece of litmus paper between the two halves, press them together and again put the ball in his pocket. When we had returned to the barn, he would take the balls out of his pocket, break them, and examine the litmus paper to see how much limestone was needed.

I commented on his mud balls. He said, "I learned that in high school. You know that you can learn a lot about a soil if you make mud balls with it. You see this soil is a good loam soil. It isn't sticky. When I broke open the ball, I noticed that it was crumbly. If you smell these balls when you break them open, they can tell you a good deal. This one has a nice, clean, earthy odor. That indicates a well-aerated soil. Sometimes, when you first break mud balls open, they smell like a cesspool. When that happens, you had better check on drainage."

I thought that I was the only one who had respect for mud balls. I didn't tell him that I, too, had studied mud balls. There are many things that you can find out about a soil in the field without taking samples to a laboratory. I feel sometimes that I can make good recommendations in the field without a soil test; but a soil test does give one more assurance.

There is a very serious problem in some of our sandy loam and silt loam soils which indirectly is associated with lime deficiency. As the clay portion of these soils becomes deficient in calcium, the particles tend to hold more water in chemical combination. This tends to make the chemically active clay slippery, so that it actually shifts from the surface layers to the layers below, or collects at the plow depth, because this portion of the soil is not disturbed in our farming operations. Over the years, with no lime added to the soil, this layer, where clay is impregnated into the silt and gravel, becomes 2 to 4 inches thick. It becomes very acid, which prevents roots from penetrating. It often becomes so dense that it prevents water from moving up or down and nutrients from moving to the surface. It is referred to as a plow sole. Every time the ground is plowed, the plow slides over this layer and tends to seal it just as a mason's trowel seals the surface of concrete. When it is wet, it becomes putty-like and, when it is dry, it bakes to the hardness of a brick. When it is dry, the plow won't penetrate it. When it is wet, the plow will bring some to the surface. Farmers will tell you that if you turn very much of this "yellow clay" up to the surface, you will sterilize the soil. It is very acid, and unless it is thoroughly mixed with the other soil (something that cannot be accomplished with a disc harrow), seed sown in it won't germinate. However, if 2 or 3 tons of limestone is applied after plowing and mixed in with a springtooth harrow, this clay becomes saturated with calcium and will tend to increase the yielding capacity of that soil. If you test the plowed layer, you may find a test indicating a very fertile soil, and yet you may not harvest 30 bushels of corn from the soil.

Soils having these plow soles will produce fairly good yields if moisture is ample. All the roots are located in the surface soil and, since moisture can't move to the surface, the only moisture available to the crop must come from rain. Plants grown on such soils will wilt during hot days and will burn up in dry spells. Trees, shrubs, lawn grasses will not do well. They have shallow roots. Some scientists have classified plants according to depth of rooting. It doesn't make sense to me. All plants will root deep if they have a chance.

The need for irrigation becomes urgent on such soils, and it is apparent that irrigation equipment can be justified. Many of our vegetable crops along the Eastern seaboard are grown on such soils, and irrigation equipment is very much in evidence. Without water, crops are not good on such soils.

A simple solution is to plow under large quantities of pulverized limestone and mix some with the soil on top. If enough limestone is applied, the correction will take place in a year. A subsoiler will speed up the action. As soon as the limestone begins to penetrate the plow sole, roots from crop plants and weeds will begin to penetrate and will gradually make the plow sole porous. If there is sufficient limestone present, the process of correction is fairly rapid and permanent.

Pond holes or depressions in a field where water tends to stand after a heavy rain are caused by these plow soles. Water from surrounding areas picks up small amounts of clay over the years and, as it accumulates, it gradually drops its clay. This clay is usually unsaturated as far as calcium is concerned. This will settle down and virtually plug the soil. Some of these pond areas, when first drained, will not grow crops because the seed cannot germinate in a clay that is only partially saturated with calcium.

I have corrected many of these pond holes and in so doing have greatly enhanced the value of the farm. My method of correcting these pond holes is to circle the ponds four or five times with a subsoiler which is set to penetrate 16 to 20 inches. This dries the pond, because any water flowing toward the pond penetrates into the soil before it gets to the pond. After the soil in the pond is dry enough to handle, I spread 4 to 8 tons of limestone per acre over the bottom as well as the surrounding soil. Then, I subsoil the field and pond area. This subsoiling should be done when the subsoil is dry, so that it shatters the soil rather than merely slicing it. Another advantage is the fact that loose, dry soil on the surface along with limestone will drop into the subsoiler furrow, carrying some limestone to the depth of the subsoiler furrow. On some occasions, I have mounted a fertilizer hopper on the top of the subsoiler and dropped pulverized limestone into the furrow.

I had occasion to work with a grower in eastern Virginia who owns a farm in the Portsmouth–Coxville Bladen soil area. These soils have a heavy, black clay subsoil through which water penetrates slowly. It was not over twelve feet above sea level and within a few miles of the Atlantic Ocean. The neighbors told me that this particular farm had been cultivated every year, but in forty years that they knew it, had never grown a profitable crop. We spread 6 to 8 tons of pulverized limestone on the surface and worked it into the soil, leaving a 10-inch surface layer of loose, friable soil. Then we mounted a distributor on a subsoiler so that we could drop limestone into the subsoil. The subsoiler was set to penetrate 22 inches. We calculated that we dropped 1½ to 2 tons of limestone into the subsoil. Much of this top soil was also dropped into the subsoil because it was so dry. The following spring this field was planted to corn and produced over 100 bushels an acre.

In this case the problem was not a plow sole but an unsaturated condition in a soil that had a very high lime requirement. Getting the limestone into these subsoiler furrows, which were crisscrossed 3 feet apart each way, helped the roots penetrate to the bottom of the subsoiler points.

When a farmer thinks about expanding his yield, he immediately plans on adding more land. A few have found out that it is more economical to cultivate the farm below the one they have—the second 6 inches of land. It is amazing how many of our crops are being grown in 6 inches of soil. That is about as deep as roots may be found in the soil, and in most cases the moisture the crop has to grow on is what is found in that 6 inches of soil.

Some of our growers (too few for our own good) have found that by preparing the ground to a depth of 12 inches, they have doubled their yields with very little more cost than it took formerly. Much of our coastal plain land has been plowed shallow for so many years that it is difficult to plow deeper. I find the same true in much of the Mississippi Valley, where appreciable amounts of fertilizer have been used. *Plowman's Folly* says that you don't have to plow and states further that clay turned up will ruin your soil. Both are lame excuses. The shortest path to the poorhouse that I know of for the vegetable grower and general farmer is not to plow coastal plain soil in its present condition. The damage done by clay that is turned up by deeper plowing can easily be corrected with lime. Furthermore, shallow-plowed land can be plowed deeper, but it takes more power, slows down operations and, in some cases, it may take special equipment. There is an easy way out. If liming material is plowed under year after year, the lime will sweeten the subsoil and gradually soften it so that the plow can penetrate a half inch or so deeper every year. By doing this for several years, the surface layer will become deeper and the clay that is turned up can be mixed with the surface soil that it came from originally. This will be highly beneficial to crop production, since it adds to the base exchange capacity of the surface soil.

The T.N.T. plow should be more popular with growers on sandy soils. It breaks up the subsoil without turning up much of the clay on top. But this plow should be used in combination with a good liming program. The use of a subsoiler may be necessary in some cases. We had a bad spot in one of our fields where the T.N.T. (Oliver) could not penetrate. We used the subsoiler first and then were able to break the subsoil with the T.N.T. plow. It makes our farming practice more efficient.

It is possible to make money with low yields at high prices, but high prices mean reduced consumption. Low yields at low prices lose money for the grower. Too many of our growers think that if they can't make money on 200 acres they need more land to get them out of the red. Some of our growers may think this is a ridiculous statement on my part, but too often one hears the statement made, "I don't have enough land to make money." Prices also come in for much discussion. I once heard a grower say, "We have to get together and set up minimum prices." His yields were low and he was losing money. I convinced him that he must increase his yield by putting on enough lime to satisfy the needs of the soil. His yields increased 500 per cent. Several years later I asked him whether he still thought there should be a minimum price on crops. He said, "No. With production I can give the stuff away and still pay my bills."

The old saying "Nothing tried, nothing gained," should be

printed on a card, framed, and hung on the back door where it may be seen every time a person walks out in his yard. If we don't try something new, we never learn. People who have seemingly good reasons get in a rut. It is easier to do things the same way year after year than to try something different. We make progress by trying new things. That is the way civilization has come about. Just remember this: the largest yield has not been produced. There is always room for betterment. Perhaps these practices won't work on your farm, but you won't know until you try them. The minute a person says, "That won't work," he has dug his rut a little deeper, and finally gets in so deep he can't see over the edge. Then his usefulness to society and to himself is lost.

A grower told me he would never use another ton of lime on his farm. I asked him, "Why not?" He pointed to a field and said, "I put a ton to the acre on that field and could see no difference." I said, "Did you leave a strip without lime?" "No." "Then how do you know it didn't do any good?" He didn't have a good answer. We looked at the field, dug up some of the soil. I told him that his particular soil probably should have 7 or 8 tons of lime per acre before he could get real results from it. From the way he looked at me, I know he figured I was out of my mind. I asked him if he would put some strips of lime across the field—2, 4, 6, 8, and 10 tons—to prove to him that I wasn't what he thought I was. He said he wouldn't waste his time because it wouldn't work. We have too many people in agriculture who think they know all the answers and remain in a rut because of their attitude. And yet farmers have tried this simple experiment and found they could double their 65-bushel yields.

We have tremendous possibilities for increasing our yields by deepening our soils, farming the second and third farms that we own, if we will only unlock the door that will make it possible for our crops to feed in those lower reaches. On our acid coastal plain soils, the key that will open that door is lime; enough lime to satisfy the needs of that particular soil. And when we once accomplish that, we will not only increase our yields per pound of fertilizer used, but we will produce food that will nourish us better and make us a happier people.

So, I wonder whether we actually have advanced in our knowledge beyond what the Indian knew when he buried a fish under each hill of corn. The soil testers can help you, but they are not the means to profitable yields, except in the hands of trained soil chemists.

Foliar tests have been discussed for many years. Here again, one wonders what people are trying to do. There is no question but that we are equipped with methods for running foliar tests—but for what purpose? We certainly have to superimpose our foliar tests on soil tests. We would like to test plants to find out what we should feed them to grow bigger yields. But do we know how to interpret our data? We have hard and soft plants or succulent and non-succulent plants. If we are dealing with fruit crops, what is our standard of excellence? We can tell by looking at a plant whether it is hard or soft. We have certain symptoms which most plant scientists recognize or can find out about; but to know what causes these symptoms may be more tricky. I have seen what appeared to be phosphorus deficiency corrected with limestone—and also by a rise in temperature. I have seen what appeared to be potash deficiency corrected by a broadcast application of limestone—and also by a heavy rain. This throws the whole problem back to the soil. A friend of mine found out that stalks of corn grown with certain treatments were sweet while others were sour.

Too few of us have paid much attention to the succulence of the crop and crop yields. The condition of the plant has much to do with the yield. The thing that causes oats to lodge also reduces their yield. A hard plant is high in starch, high in proteins, and very low in amino acids. A succulent plant is low in starch, low in proteins, but high in amino acids. Amino acids are bitter. Such a plant does not yield a good crop nor does it produce good quality. Feeding tests have shown this to be true. To test this plant for nitrate or ammonia may not be much help, because we have no practical way of reducing the succulence.

There are too many questions for which we have no answers. If we have a hard plant that isn't growing fast enough, we know it has a surplus of starch and it needs more nitrogen. Then the question is whether we have nitrogen or oxygen in the soil. If there seems to be enough nitrogen in the soil, perhaps the plant can't assimilate the nitrogen because of a lack of calcium in the soil. Perhaps we have lice on the roots. Perhaps we have some other toxic materials affecting the roots. Foliar tests under these conditions are primarily of academic interest. My candid opinion is that if we pay more attention to the degree of calcium saturation of the base exchange complex, we will have very little need for foliar tests—unless we are interested in the level of nutrients for academic reasons.

CHAPTER 14

The Farmer and His Profession

No other human occupation opens so wide a field for the profitable and agreeable combination of labor with cultivated thought as agriculture. Every blade of grass is a study; and to produce two where there was but one is both a profit and a pleasure. The thought recurs that education—cultivated thought—can best be combined with agricultural labor, or any labor, on the principle of thorough work, and ere long the most valuable of all arts will be the art of deriving a comfortable subsistence from the smallest area of soil.—ABRAHAM LINCOLN

THE PROBLEMS confronting the farmer are partly his own fault and partly due to economic conditions beyond his control. One problem which he has not been able to solve—one which no one else has been able to solve—is how best to adapt his farming operations to the kind of weather he can expect. So far, he has had little control over the price he receives for the produce he grows. He can, of course, buy land in any location and be assured of a certain climate. Therefore, he has no idea what his yields or gross income will be and whether he will receive enough money to pay the costs of preparing the crops for market. If he cannot realize a profit when his crop has been marketed, he may work a whole season for nothing. This has a demoralizing effect and has made his profession a gamble. He must invest anywhere from $5 to $200 an acre to produce a crop before he knows what his acre return will be.

One of the big problems in establishing a price that will give him a profit above his costs is the variability in yields on different farms which, of course, cause unit costs to vary within wide limits. The ability of people to manage affairs greatly influences costs.

A farmer has no way of knowing whether he will lose his crop because of a drought, a flood, a wind, or hailstorm. An early fall or late spring freeze can ruin him. His crops may be destroyed by an avalanche of insects or some plant disease. Or, because of unfavorable weather conditions making it impossible to cultivate his crops, he may see weeds reduce his yields to the point where the crop is not worth harvesting.

As a generality, cattle farming and poultry farming have been a little surer than all-grain crop farming, because pastures are not affected as much by the elements as other crops. However, cattle or poultry farming have not been a bed of roses either, because they too are dependent on adequate yields of crops and prices of milk, eggs, and beef.

There is a wide difference in the amount of supervision farmers must expend on their farms. A dairy farmer is probably busy the year round. A poultry farmer can be busy for twelve months out of the year. A beef-cattle grower and feeder probably is busy ten months, unless he feeds his cattle on pasture—in which case he may have several months a year for vacation. A grain farmer spends only three to six months in actually farming. A potato grower only farms four or five months out of the year. Storing and marketing the crop may add two months. A vegetable grower can easily spend nine months out of the year supervising his crop. A fruit grower may spend from six to nine months supervising his operations. When we consider any crop program, we must realize that the growing season from early spring frost to late fall freeze determines which months will be busy.

Thus, the farmer figures how much money he makes per month while he is busy. Certainly, the potato grower cannot expect to make twelve months' pay for four months' work. If he can make good wages for twelve months by working four months he certainly is in a good profession.

Statistical portrayal of total crop production in the United

States has led our uninformed proponents to paint the farm picture a rosy industry. Such data is not proven in terms of profits and, as a result, a four-billion-dollar grain crop sounds like big money to the farmer; but when it costs four billion dollars to get that much money in return, it has meant no actual profit to the farmer. A farmer's pay must be figured on his hourly wage, which includes his profit. He must not overlook the fringe benefits, such as rent and produce.

On the basis of actual profits, agriculture is not a going concern for the average farmer. Any business run on the same basis would be bankrupt. Too often we have seen profits in one year wiped out by two or more years of less than cost prices. Farmers have lost their farms because of fertilizer bills, or by assuming mortgages for needed buildings and not being able to make sufficient profits to even pay the interest.

Forty-odd years ago a farmer could hew himself a farm out of standing timber, erect his buildings, and by hard work build up an equity from $5-an-acre land into $100- to $200-an-acre land in ten years. My father took 80 acres of land which cost him $400 ($5 an acre) and in seven years, cleared it, erected barns, silos, machine sheds, and a house and sold it for $12,000. The $12,000 was clear. He owed no money. I doubt whether a person could do that today, because of the high costs of things he has to buy. I remember that the first large barn, 34 feet by 60 feet, that Father built cost him $800 in actual cash. The following year he raised 5 acres of potatoes and paid off the $800 mortgage because potatoes happened to sell for $1 to $2 a bushel. Today that same barn built in the same community would cost $4,000 and it would take the profits from 160 acres of potatoes to pay for it. That comparison shows the reason why the farmer is in trouble today and it is something that he can do very little about.

We hear much about parity financing between industry and farming. It is not well understood. Parity assumes that a man growing farm produce should realize as high an hourly wage as the man who runs a manufacturing plant with equal capital investment. This is based on average farm income for a previous 10-

year period. However, our economists have overlooked the fact that a $6,000-a-year farm income may have been obtained with the help of three grown children and the man's wife. Many a farmer's wife has spent many days driving a tractor, tending chickens, and milking cows while the farmer was doing other necessary work. I have seen ten- to fifteen-year-old boys and girls drive a tractor all day; and no accounting was made for this share of the $6,000 income which the farmer received. These things rarely happen in industry.

The incentive that got young people started as farmers forty years ago is no longer here. With conditions as they are today a person does not dare to start life on a farm, unless he has the money to buy the land and an equal amount to buy equipment. He does not dare to assume the responsibility of a mortgage because he has absolutely no assurance that he can make enough profit to pay it off. Young married women don't put up with the inconveniences that my mother did when she and my dad bought wild land and with little cash built a farm out of it. Some people who read this will contradict me and say it can be done, and I will have to agree that it is being done by a few hard-working, experienced individuals who possess the love of accomplishment. But for every one who can do it, there are one hundred or more who could not make the grade.

Farm life should be the most attractive and healthful vocation in the world, and yet we know it isn't. Young people don't want to stay on the farm for various and sundry reasons; and the main reason is that they cannot have the luxuries and conveniences of city life. Many want better education and, having gotten it, now are attracted to the salaried jobs. Farming has not paid enough to put in running water and electricity so that they could enjoy the privilege of a shower or tub bath and toilet facilities that would eliminate the outhouse in the back yard. They want to see movies and have the use of an automobile. Today television has partially filled that need. The farmer who has been able to provide these facilities is in the upper 10 per cent of the profession.

Many estates that were built in farming areas were constructed through cheap labor, slaves, or children. The farmer with a large family was fortunate, because the help he had cost him only their keep.

As a result of economic and social changes which have occurred during the past fifty years, we have people farming who either love the land, have their life earnings invested and can't liquidate their investment, or are incapable of doing anything else and live on the land to glean enough food to keep from going hungry. They may use the farm as a place to live while they pursue a salaried job; or they may have accumulated enough wealth to farm in spite of poor prices. We have some large farming corporations, some of which through sound business principles, are able to make a good income while others usually operate in the red.

There are very few, however, who farm because it is a sound business enterprise. This is unfortunate, because every human being in the world is dependent on the farmer for his food. Manufacturing is dependent on the farmer for raw materials. As a nation, we must protect ourselves by protecting the farmer. We must furnish an incentive that will make people want to farm as a vocation. There are enough people who would farm if they could earn enough to give them the luxuries that they enjoy on a salary. In other words, we, as members of a society, must take sufficient interest in our agricultural industry to assure the farmer of profits sufficient for him to live as well as his city cousin. Unless we do this we may find it impossible to buy sufficient food to nourish our bodies. We have seen this happen in China, in India, and in Russia. We can do something about it if we will. Some claim that we have reached the point where, as a world population, we do not have sufficient land to grow food to feed us. The picture is not as dark as some think, but we still have to change our economic picture considerably. It has been said that our civilization has reached the crest and that we are now on the toboggan headed for oblivion. How can we slow this descent and remain on the crest? By solving the farm problem we can return to an even keel and stay there for centuries.

There is much being written today about our diminishing food supply, our ever-increasing population, and how long our soil resources will continue to supply us with food to satisfy our needs as well as a surplus to export to food-hungry nations. On the basis of what has been published during the past 150 years, the agricultural industry should be the most prosperous of any of our industries; yet the average farmer, with his low yields, can't depend on the price set by supply and demand to pay his costs of production. Subsidies from the government, because we have surpluses, have been tried, but help little.

We talk about our diminishing land per capita. The fact that our population has been increasing for several hundred years while we remain capable of producing more food and more raw materials than we can use should be a warning that perhaps our thinking has been running on the side track rather than the main line. Abe Lincoln asked several questions that we may ask ourselves in all walks of life: "Do we know where we are?" "Do we know what we want?" "Do we know how to get what we want?"

We have had government programs to help agriculture, but none of them has actually helped to bring order out of chaos. I have heard the statement made that we need a war to restore our agriculture to a prosperous level, because a brisk demand for food raises the prices to the producer. It is true that farm prices have increased during war periods because the demand was high and the supply was just a little below demand. It was so much so that we had to resort to rationing and impose a ceiling on prices so that the consumer could buy enough food. I often wonder why we need ceilings. A person has a certain amount of money to spend. If he has to pay too much for some items he will buy a cheaper item or a cheaper substitute.

If we are going to impose ceiling prices we must also have support prices. In other words, we must subsidize agriculture. This may not be sound philosophy, but it is necessary to encourage our farmers to produce more food, and this encouragement can only come as a result of sufficiently high prices, so that the farmer can depend on a profit year after year. But why should we have to do this? Why don't we subsidize other industries? Is it because the farmer has the key to maintenance of life in his hand? Perhaps Daniel Webster wrote words of real wisdom when he said, "If there is one lesson of history that is unmistakable, it is that material strength lies very near the soil."

We have supported our agriculture by appropriating money for research to a higher degree than any other country. We have done this in good faith and with a certain amount of clear thinking, but I wonder whether, as Lincoln intimated, we always knew what we wanted before we appropriated money. Funds appropriated as a result of political pressure are not always spent wisely. There is much waste and duplication and, as a result, we have burdened agriculture with something which is always open to criticism by pressure groups. However, the agricultural industry in the United States has fared well from the standpoint of receiving its share of governmental money. Perhaps it is one reason why people in the United States are well supplied with food.

The fact that the Armed Forces found much malnutrition among the draftees is no reflection on the ability of our farmers to feed them. That is a problem for our nutritionists and sociologists to solve and plays no part in our discussion. Our problem, from the point of view of food supply, is to decide whether we have produced enough food, how much longer can we continue to produce enough, and what means should be adopted to stimulate the production of more food.

According to the law of supply and demand, which functioned to a much higher degree in the early part of the century, agriculture should prosper. But it hasn't; perhaps we should decide why. Perhaps it doesn't apply to agriculture, in which case we make a mistake in assuming it guarantees an adequate food supply. There are several things worthy of consideration.

Has government interference, in the form of special taxes, programs of curtailment, price ceiling, price controls, or other acts, had some deleterious effect? Has weather anything to do with it? Has the cost of labor anything to do with it? There are many angles, which should be studied from different viewpoints. A cursory view of these factors, I believe, shows that there are too many obstacles to permit the law of supply and demand to take care of our food supply. We can take an optimistic view of our ability to feed our people for several centuries to come; but it will require some planning based on clear thinking. The question is, can we set up a world program that will prevent starvation and famine?

We do have famines in some countries. India and China have had famines. European countries have been more fortunate. It is interesting to consider them. India and China have been pointed out as places where the population exceeded the capacity of the land to feed them. Is the fault in the soil or in the government? India has a far greater population per square mile than the United States. However, I am convinced that we in the United States could easily feed as many people as India has per square mile. I can't prove this. But there are many reasons why this might be so.

Transportation in India is not adequate to transport food from areas of plenty to areas of famine and, as a result, only wheat and other grains can be depended on, because they can be stored. However, grain must be grown in suitable areas, and a drought can easily wipe out a crop. Perishable crops must be grown close to the centers of population unless good transportation is available. India has insufficient industry. When a nation's population is largely agricultural, cultivating land to produce food becomes a hand-labor job, which relegates that part of the population to the peasant class. Each individual may provide for himself, but has little left for his city cousin.

Indian agriculture has been given no support by the government. There has been no encouragement to produce food. Agricultural pursuits have not been marked by scientific methods of research. Again we come to the possibility that a lack of land may not be the real cause.

What is said of India is also true of China. The fault goes back to Daniel Webster's quotation. The Old World idea was that the farmer was of the common class, fundamentally necessary to the world but not capable of entering into the governing body. We can be heartened by the fact that our thinking in this country has gotten away from the idea that anyone who can't do anything else can be a farmer. Agriculture cannot be a dumping ground for humanity. A successful farmer today must be well trained in the sciences and of keen mind.

The Romans took a different point of view. When they wanted to commend anyone, they said the tillers of the soil made the best soldiers, were the best husbands, and were the source of state leaders. Italy has had no famines. Germany and the Low Countries looked after their agriculture. They realized that a happy, contented man was one who had enough food. They set up experimental stations to study problems of the soil to help the farmers maintain the productiveness of the land. They, too, supported their industry. The fact that a fanatic came into power and caused the downfall of Germany is beside the point in this discussion. What has happened in France and Spain is a good example of what happens when a government overlooks its agriculture and industry.

In this connection we can examine the communistic state. In Russia, we have a good example of lack of co-ordination between science and the agricultural industry. Russia is supporting the agricultural industry by research, but the machinery to get scientific facts into the hands of the farmer has become disorganized. Communal living and government farming are not conducive to a strong agriculture. We must have free enterprise among farmers as well as among industrialists. Part of the success that our farmers have enjoyed is due to the fact that they could carry on their operations with a minimum of interference.

Communal living discourages initiative. There may be some advantages for the man who can't think and has no ambition, but some other means should be provided for his welfare. This is a sociological problem which should not be heaped on the shoulders of the agricultural profession. This should be the concern of the government and should be divorced from agriculture. Such people must work under direct supervision; many of them do make up the laboring class on our farms.

Good examples of government interference are the tariff and certain types of taxes, such as the tax on butter substitutes. People in the United States will take certain regulations up to a certain point, and then they rebel. The supporters of taxes on butter substitutes will soon find that they have cut off their noses to spite their faces. The price of butter is high because the supply does not meet the demand. Furthermore, there is little indication that it ever will. The tax on oleomargarine is a spite tax rather than any help to the butter interests. As people become educated and realize that vegetable oils are as nutritious as butter the per capita consumption of butter will decrease. This will hurt the butter producers. With present costs of producing butter even high prices will not encourage the expansion of the industry.

Dairy farming is hard work. The younger generation doesn't want to be tied down to milking cows. Dairying developed because of the low price of family labor. A man with a large family makes a big profit in the dairy business and, as long as he can command his teen-agers to work for him, he will continue to make a profit. The dairy farmer will be supplanted by the fat stock man because the stock grower can get along with much less labor. The high cost of labor is going to be responsible for many changes in our agricultural pursuits. When people can make good wages in industry by working 40 hours a week they want equivalent hourly wages for the 50 to 60 hours they work on the farm. When our dairy interests realize this they will try fairer means of helping their interests than killing competition by taxation. The dairy industry will become a fluid milk business, developing closer to population centers or within feasible hauling distances of such areas.

How about the price ceilings and support prices? Price ceilings tend to discourage production for the reason that boards who sit together to set those ceiling prices are composed of those who have a high degree of efficiency in their operations. I have in mind a survey made in a community where tomatoes were produced for the fresh market. The costs of growing tomatoes for 100 farmers ranged from 40 cents to $2.25 a bushel. A good selling price would be $1.80. After a few years, those who produced with costs close to or above the selling price will discontinue growing tomatoes.

Interference by the government can be good or bad. Regulations on price-spread between the producer and the consumer would help to improve any industry. We need people who handle produce and commodities. The fact that we have so many is an indication that profits are lucrative. A fair profit which would become lucrative only by handling large volumes would be permissible. The present system of large profits for the middleman is on the way out. Too often these profits have put the producer out of business; but if we are to assure ourselves of adequate food supplies we must encourage more people to produce.

The government support price based on parity has many weaknesses. We could support ourselves right out of the picture, when it is based on parity. The 1948 support price on potatoes is a good example. By the time the potatoes reach the consumer he won't be able to pay for them and the demand for potatoes will decrease. Such practices are responsible for initiating vicious circles which do us more harm than good. If the retail price is too high the laboring man strikes for more wages. This in turn boosts prices still further. We have many groups looking for special favors who do not realize that as they demand and get more they must pay for it, whether through taxes or commodity prices. As a result we have demanded ourselves to a level where we can't do business with neighboring countries, so we set up tariffs to protect ourselves. If our neighbors had the money and could buy our agricultural surpluses, we wouldn't need support prices. There is a big demand outside of our own walls for more food than we can produce.

Democracy is a wonderful institution, but it will be short lived if we don't do something to adjust our economic life. A high standard of living is of little value to the populace if we can't enjoy prosperity. Many of us feel that we must have plenty of money to spend in order to be happy, and our aim seems to be to acquire more. As a matter of fact we probably have less money to spend now than we did fifty years ago. To be prosperous we must produce more per capita. The present system has developed a number of weaknesses and inequalities. High wages immediately start talk of inflation, but when only one group gets high wages, while the farmer and the white collar worker are underpaid, it is difficult to see why we should have inflation. If people buy only in proportion to the money they have to spend, there is little danger of inflation. Inflation comes from easy to obtain credit. This hurts agriculture, because people cannot buy the food they need. If the government wants to help agriculture, a curb on credit would do more good than a support price. A person tied down with heavy monthly bills cannot buy high-priced food. This is partly the reason why we have malnutrition and starvation in the midst of plenty. We need closer co-ordination between industry and agriculture. The only agency that can do this is our federal government, but any program should be the product of thinkers, not politicians. It is not the job of politicians to establish a program. It is their job to put it into effect after it is agreed upon.

Furthermore, the world situation must be taken into consideration. This is more difficult, not so much from the standpoint of European as from Asiatic countries. It may be that subsidies to foreign countries may be a better method of world distribution for our surpluses than subsidizing crops in this country. Our standard of living is the result of our wage scale in relation to that enjoyed by our foreign neighbors. The upper limit of our wage level is responsible for the rapid advance in the adoption of automation in our factories which, of course, is increasing the ranks of the unemployed. Industry is adopting automation to keep solvent. Thus, the seesaw between the survival of industry and labor unions continues to increase prices. This creates friction not only between industry and agriculture, but between the United States and foreign countries. Thus, if we would carry on trade with our foreign neighbors, we must subsidize our agriculture or loan money to our foreign neighbors. We do have the possibility of shortages in this country and I would much rather have a surplus every year than control acreage too closely. Any tax for surplus produce would be a justifiable burden on society in general. The question then is: How can the farmer produce that additional yield per acre and how will that affect our national economy? The answer to that question involves a number of considerations.

In general it is safe to say that the average American farmer has at his disposal information which would enable him to farm much better than he does. Experiment stations the country over must advise him how to increase his yields considerably above the average, even though they may not have the information to tell the good grower how to achieve his maximum yield, which is two or three times the average yield. If we analyze the situation, we can, for simplicity, say that one-third of the farmers produce below average yields, one-third produce average yields, and one-third produce above average yields. This varies with localities. There are isolated areas which support higher average yields than others. A good example is the California potato growing area, which produces 300 bags, whereas the area in Tidewater, Virginia, produces an average of less than 100 bags. And yet there are growers in Tidewater who produce on occasion 350 bags per acre just as there are growers in California who produce 500 bags per acre. Thus, soil conditions have a big influence on potato yields.

It has been my privilege to see 150 bushels of corn per acre harvested where the average yield is 30 bushels; 1,000 bushels of spinach per acre where the average is 230 bushels; 85 bushels of oats where the average is 27 bushels and 504 bushels of sweet potatoes where the average is 83 bushels an acre. The first thing that comes to one's mind is that better soils were responsible, but in not one of the cases mentioned was this true. It was due to superior cultural conditions. It convinces me that good yields can be produced on all our farmlands, even though they may not always be profitable yields. One thing that we can be sure of is that a very small percentage of our growers grow as good a crop as is possible for the prevailing weather conditions in a given area.

Our crops are grown from water and sunshine, with a little fertilizer and lime to make it possible for the plant to make use of the sunshine that it receives. A corn plant contains 80 per cent water and 20 per cent dry matter, of which less than one per cent is ash or lime and fertilizer that was applied to the soil. The remainder of the dry matter is made by the action of synthesizing sunshine and carbon dioxide in the air and moisture taken from the soil through the roots and dew taken in through the leaves. And yet this dry matter per acre is far from the quantity that could be produced. A 300-bushel spinach crop produces 720 pounds of dry matter for food, while a 50-bushel corn crop produces 3,000 pounds in grain and cob, and another 3,600 pounds in the fodder. It costs $81 to grow the spinach and $41 to grow the corn. The cash returns on the spinach could be from $225 to $600 an acre, while from corn it could be from $37 to $75 an acre. From the standpoint of food energy there is far more in corn than in spinach. Corn therefore becomes a cheap-land crop, while spinach can be grown in the shadow of Radio City. The farmer can gamble with spinach while he has to be much more conservative with corn. The man who grows 50 bushels of corn probably would not dare to risk a spinach crop. Spinach is a perishable crop that must be marketed quickly, while corn may be stored and sold at top price. In other words, we have the intensive truck grower and the extensive field crop grower, to say nothing of fruit growers, cotton, tobacco, wheat and potato growers—each growing a different type of crop, each crop requiring special handling, but each crop doing well on the same soil type, each with an adequate level of fertility and lime that must be maintained in the soil.

We can go further and say that a man's genetic background or make-up plays an important part in how successful he can be with such a crop. Each and every one has available information to do better than he has been doing with his particular crop. And yet it isn't being done. This is shown by the consistency with which our average yields remain the same from year to year.

The average yield for potatoes in 1945 was 151 bushels per acre. If that yield should happen to be increased 20 bushels per acre, it would have increased our production from 430,773,000 bushels to 486,773,490 bushels, and if it were increased to 200 bushels per acre it would produce a total crop of 538,466,250 bushels. This could happen but probably won't, because many factors control our yields and farmers are not all equally good potato growers. Prices will be determined by supply and demand. Our total crop, to a certain extent, will fluctuate. Acre yields, which determine the number of dollars that a farmer makes, are affected by weather conditions, prevalence of insects and disease, amount of fertilizer applied, and soil conditions. The farmer who gets those big yields knows how to handle each one and sees that practices are followed through. So, we have not only a cultural problem but an education problem. How far we can go with our educational problem is a question. If we use average yields as our criteria, progress will be slow.

There is another trend that is being felt. The per capita consumption of potatoes is decreasing, giving way to other truck crops. There is less need for high-energy foods. The proportion of white collar people to laborers is increasing. The demand for high-energy foods is decreasing while the demand for high-vitamin foods is increasing. The lettuce-tomato sandwich is gradually taking the place of the potato and gravy diet. This has a disturbing influence on our crop producing areas. Farmers must stay on their toes if they would keep abreast of the times.

CHAPTER 15

Plant, Animal and Human Nutrition and a Proposed Fertilizer Program

THE NUTRITIONAL NEEDS of human bodies are only partly dependent on the food we eat. Heredity probably plays a major role in how well we can get along on the food available to us. We know that people vary in their allergies. A given food may be good for some people and not for others. If we check on the nutritional needs of people like the Eskimo, who lives on fish and blubber, and then read a treatise on what a well-known nutritionist tells us we should eat to be healthy, we begin to wonder whether our thinking is as sound as it might be. The Eskimo lives in a cold climate. The man in the tropics lives under high temperatures where energy values are less important. And then we have all intermediate areas. We find wide variations among people. We have the thins, the fats, the talls, and the shorts. All probably require vitamins and minerals in widely different amounts. But regardless of what we do in the agricultural field, ultimately we get involved in human and animal nutrition.

As a boy on the farm, I heard about "easy keepers" and "hard keepers" among horses. I have observed that we have people who fit into similar groups. Thus, if we assume that heredity controls sizes, shapes, and so on, we probably have to assume that for any one individual we may have short fats, short leans, tall fats, and tall leans, all of whom have vitamin and mineral requirements. This seems to be true in the tropics as well as in the frigid zones. And the ease with which people build up body weight undoubtedly is the result of how many calories they take into their bodies. We know that when a person stops eating, he loses weight, while a glutton usually is heavy. But, experimentally, we must learn by trial and error, since we have no two individuals possessing the same heredity.

How well a person feels depends on how well his glands function, which also involves heredity, and the minerals and vitamins he consumes. Because of the law of "survival of the fittest," people have become more or less adapted to their environment. Those who don't fit in die young.

Nutrition of humans is closely tied in with soil conditions. Vitamins and minerals undoubtedly have considerable bearing on how well people feel, but this is beside the point, whether one takes on weight or not. The food we eat comes from the soil, whether we eat the seed and foliage of plants or whether we eat meat. Food from the sea is considerably different, perhaps much better for us than a beef steak from a steer grown in a feed lot with a corn diet. In other words, even while we admit that there are similarities in the way humans and animals use the food they eat, we must assume that basically our nutrition depends on what minerals are available in the soil and how much sunshine our food crops receive while they are growing.

Lately, it has come to our attention that the palatability of our food that we grow for our animals depends on the amount and kind of fertilizer we apply to the soil to grow the crop. There is also good evidence that the manner in which the crop is grown and fertilized determines how many pounds of corn silage is necessary to produce a pound of beef. Apparently, the seasonal weather conditions, water, nitrogen, sunshine, and general fertility level determine the nutritional value of the crop. Conditions favoring rapid growth produce proteins and starches as well as other similar products.

Protein, a term generally applied to certain compounds, is the result of nitrogen, starches, and sugars being combined through chemical reactions in the plant supported by sunshine. Amino acids are an intermediate stage. The amino acids are water soluble and are the building blocks of the proteins. In the process of condensation, water is removed, and the final storage protein becomes insoluble in water but retains certain chemical properties which can affect the growth of the plant.

An amino acid is water soluble and very chemically active, but contains comparatively small amounts of caloric energy. It contains nitrogen, carbon, hydrogen and oxygen in a hydrated form.

A protein (proper) is insoluble in water, usually stored in the plant for future use, and has considerable caloric value as a source of energy for the production of meat. Proteins also contain nitrogen, carbon, hydrogen, and oxygen, but in a dehydrated condition. The ideal food for animals is, of course, a proper balance between starch—which the plant makes in its leaves as a result of the carbon dioxide absorbed by the leaves and water, with the help of the all-important sunshine—and proteins. Part of this starch is used in growth, and part is the surplus that is stored after the plant has used what it needs to combine with the nitrogen taken in through the roots to form the proteins. If there isn't enough starch made in the leaves, much of the protein exists as soluble amino acids. If there is a surplus of starch, then more of the energy-filled storage proteins is deposited.

Seed, potato tubers, bulbs, and other storage organs depend for their size on the amount of surplus starch and storage protein that the plant can accumulate. Corn seed may contain 8 to 14 per cent protein, and almost 70 per cent starch or starch-like material. No. 2 dry corn should not have over 14 per cent water. The actual mineral content (phosphorus, potash, calcium, and other minerals) accounts for less than 2 per cent of the weight. In other words, when we feed or sell a bushel of corn which weighs 56 pounds, we are selling approximately 6.7 pounds of protein, of which 1 pound is actual nitrogen, 41 pounds of starch and other carbohydrates including some sugar made from the air and, at the most, 1 pound of minerals. The water content in this case would be 7.3 pounds. These figures vary according to the season and the amount of nitrogen the plant has access to.

We must remember that this corn that is saleable is surplus and is storage material. Our problem is to grow corn in such a way that the plant will produce surplus starch so it can produce large ears with heavy kernels. This gives us corn that will make it possible for the animal to produce the most meat for the least amount of feed. It also stores well in a crib and maintains a constant water supply, which prevents corn from molding in storage.

What happens when the plant does not produce enough sugar and starch in the leaves to give the maximum yield? One of the obvious symptoms is the appearance of barren stalks—stalks with no ears on them. Such stalks are large, leafy, often purplish-green in color, because the plant does not have sufficient phosphorus. Too much nitrogen available in the soil causes phosphorus to become deficient. This condition produces a plant with a large part of its protein in the amino acid form. True, it is a high protein plant and farmers are told protein is valuable feed, but it is not as valuable as storage protein. The only people who propose the use of more nitrogen are the people who want to sell it. Actually it is cutting the farmer's yield and raising his costs. There is no rhyme or reason to this philosophy. Many of our experiment station people advocate this program. Either they don't know what they are talking about or they have sold out to the nitrogen interests. We will always have plenty of nitrogen because the raw materials are free and the cost of manufacture is low. Under such conditions sales pressure will always be exerted and some of this undoubtedly blows over the heads of some of our research people. During the First World War, we had "laughing gas" shot at our soldiers to dull their senses so they didn't know what they were doing. It was made from nitrogen. I wonder sometimes whether some of this may not be mixed with our fertilized nitrogen to bewilder some of our research testers.

There are other reasons why our crops do not have the best quality. Generally speaking, a soil that has the amount of calcium prescribed by the active clay and organic matter it contains produces the most nutritious food. Actually, the available calcium in the soil pretty much determines the quality of the crop, regardless of the fertilizer treatment. With adequate amounts of lime we can make few mistakes; but without adequate amounts of calcium, almost anything we do can be a mistake.

The effect of too much nitrogen on a soil well supplied with calcium may not be serious. On a soil with too little calcium, it may be tragic. The amount of rainfall and the amount of cloudy weather can be ruinous. For this reason, the most favorable approach would be to concentrate on supplying sufficient calcium to get the physical and chemical condition of the soil into top form. Having accomplished this, only then are we ready to consider other "growth promoters" and "yield increasers."

It is true that one can do certain things to a crop to increase its growth. However, it seems rather foolish to feed a lawn, for instance, with an abundance of nitrogen when the chemical and physical condition of the soil is faulty. This will only make it necessary to mow the grass three times a week instead of once every ten days; and then the lawn burns up in August. That is not good treatment of a lawn; as a matter of fact, it is about as good a method to kill out a lawn as I know of. Very rapid top growth means poor root growth. A plant must make surplus starch in the leaves to make good root growth. If you cut off the leaves you don't have surplus starch, so you don't have good roots.

Consider Canada thistle or quack grass. Both of these plants are hard to kill because they have underground storage roots or stems in which the plant stores protein and starch. We call this storage material root reserves. The principle of killing these plants is to starve the roots. Anything you can do to prevent the plant from storing proteins and starch in these underground stems will gradually kill it. If you have a bad infestation you can kill it by fertilizing heavily with nitrogen and mowing off the tops every week. Weed killers like 2-4-D will kill thistles by causing the plants to use up these root reserves. We also have materials which will affect quack grass in a similar manner. Quack grass won't grow in a lawn because it doesn't have time to store up root reserves, since the leaves are cut off so often. It is all based on the physiology of the plant. The more we know about plant physiology, the better equipped we are to know what to do either to promote bigger yields or to reverse our method if we want to kill the crop. Any weed killer that only burns off the leaves is only effective if we make repeated applications. Too often we forget to make the proper application and we condemn the material.

The nutrition of our bodies is dependent on the chemical composition of the vegetables and meats we grow on our farms. Thus, to be concerned about human nutrition means we must be concerned about animal nutrition, which in turn means a thorough understanding of soil reactions. In any program of betterment of human beings we must start from the bottom and work up. If we can handle our soils properly, our animals will thrive better and humans will have fewer miseries. This statement is not without considerable proof. To find such proof, we must scan medical journals, plant-science and soil-science literature, farm experiences, and actual field plot experiments.

We have many people working on and doing research in all fields. Most of us are working in cubbyholes by ourselves. Often we feel our field is the only important one to consider. We don't know what someone else is doing. I am only interested in the overall picture. I do not know enough about medicine to comment on it. However, I am of the opinion that if our medical profession had a better understanding of plants and soils, many of our complex problems would respond to simple treatment.

I have been interested in the effect of the calcium ion on the growth of plants. I have seen the drastic effects of insufficient calcium. I have seen plants become stunted and actually disintegrate because of lack of calcium. I have seen what I am sure was calcium deficiency causing rotting of human flesh. I assume this was calcium deficiency because this terrible condition disappeared when the patient was daily fed 30 milligrams of calcium gluconate. I realize this is no proof, but when one sees this happening often enough one begins to feel his observations are more than coincidence. Furthermore, there are medical men who agree with me that there are certain relationships which in the popular vernacular are cures for certain conditions. Since you can't prove anything with research on human beings, you can deduce from "cause and effect observations" that at least you may be on the right track.

As a result of determining the available calcium in thousands of soils in many parts of the United States, I find there is a paucity of calcium. Crop yields have been correlated with these calcium readings. As a result of this I am convinced that most soils having over one-half per cent active organic matter must have about 2,800 pounds of available calcium, using a weak extracting solution. The following figures give some idea how this shapes up:

Available Calcium in Top 2 Feet of Soil	*Yield of No. 2 Corn*	*Yield of Soybeans*
Less than 400 lb.	8 to 10 bu.	7 bu.
400 to 1,200	10 to 30	7 to 14
1,200 to 2,400	30 to 65	14 to 26
2,400 to 2,800	65 to 100	26 to 40
Over 2,800	100 to 165	40 to 58

If we go back and scan the research results from animal feeding, we get the impression that animals with insufficient calcium become irritable, develop sores, have difficulty raising young—not unlike many of the miseries claimed by human beings.

In other words, we have more exact, proven facts about the health of our plants and animals than the medical profession has about human health, because one can't have checks to compare experimental results in humans.

If we can believe a small fraction of what we read about human nutrition, we must draw conclusions from large numbers of people: 100 in one group against 100 treated in another group based on experience from observations that consider the variability of the human race. Minerals and vitamins apparently are equally important to man and animals. From observations of their effect on animals, we can assume that in a similar manner they may affect humans.

I mention these things because we believe that as a result of our research program, it is possible to grow good crops. Good crops should be good food for our animals and the meat they produce should be the best food we can get. If we can grow big acre yields by having the minerals in our soils in the right proportions, we are contributing to the production of food that will help us to maintain a high level of good health.

It has been my honest opinion that sickness and misery experienced by humans is a reflection of what they eat. Either we don't eat the correct foods, we don't select a wide enough variety, or our foods are grown on such poor soils that they are not giving us the nutrition we need. The method or program presented in the following pages has been geared to grow food as good as I know how to grow.

Many people write about human health, diets, vitamins, and minerals. Most of these books are written in popular language by members of our medical profession and are directed to the layman because they encourage a large number of people to read them. As a result, many of them become best-sellers. Whether they do any good is anyone's guess. Some writers criticize them as worthless; others praise them. Some give a résumé of their own experiences after practicing medicine for twenty-five to forty years. Their experiences probably are worth more to the reading public than trying to figure out the meaning of many of the experiments in human nutrition. I am listing several of these books and hope you will read them, not because I feel they are authoritative nor because they have the last word on the subject of human health, but because they are all trying to arrive at the utopia of perfect health. They do all seem to have some bearing on our program of growing crops, since they all cite our poor soils as the cause of much misery. Better crops from our soils mean better feed for our animals and better food for human beings.

Folk Medicine was written by Dr. J. C. Jarvis, of Vermont. He gives his experience dealing with the health of rural people in Vermont. It represents forty years of practice. It is simply written. He deals with old, homespun remedies found in the kitchen. His stand-by is a honey and cider-vinegar mixture which some people claim has done them much good. Whether this is real or psychological is immaterial. He believes in well-grown fruit and vegetables but shies away from calcium, a mineral which I deem very important in our diet.

Eighty-year-old Doctor's Secrets, by Dr. William Brady, was written by a practicing physician in Penn Yan, New York. He was also a columnist for many newspapers for forty years. Experience makes up the background for this book. He emphasizes the need for adequate calcium in the diet, contradicting Dr. Jarvis.

Overfed But Undernourished, by Dr. H. Curtis Wood, an associate in obstetrics at the Episcopal, Stetson and Rolling Hills hospitals in Philadelphia, is written with more authority and cites many more research results. It also lists many references for further reading. This is a small book and can be read in an evening. I would recommend it for general reading.

All of these books are on nutrition, although Dr. Wood is more specific in his comments. They all bemoan the fact that our soils are becoming depleted of minerals, that our foods don't contain the minerals that we need and that therefore, to keep healthy, we must depend on vitamins and minerals along with a few other compounds.

We recognize the fact that the proper balance of minerals is very important to grow good crops. The program suggested here is trying to accomplish what these doctors state is the weakness of our whole food-producing machinery. On the basis of experiences people have had, we feel that adherence to this proposed program will help not only to build up our soil but will greatly improve the quality of our food.

I have been associated with fertilizer research work in experiment stations for some twenty-five years. My college research work was in plant nutrition. The program I am now advocating is the result of trying to add some rhyme and reason to the use of commercial fertilizer. My ideas are radically different only because by changing my ideas I was able to give farmers help which they were unable to get before. Many of the ideas I was taught in college were of little help when I came in contact with actual farm problems. I made many changes, all of which helped me to increase yields and lower costs.

After reading various books written by members of the medical profession and reading the criticisms of these books by people who had no connection with the medical profession, I realized that much of our knowledge about growing good food, building our soils to grow better food, and prescribing treatment for ailments is on a very insecure basis. I feel that a continual emphasis on the methods proposed here, backed by research work increasing yields two- to three-fold with comparatively simple treatment, will go a long way to produce more nutritious food, which in turn will result in better health.

The following suggestions are offered for trial purposes and should be considered in detail.

A SUGGESTED METHOD FOR GROWING CROPS PROFITABLY

STEP 1. After selecting land which is drainable and workable with available equipment, the profile of the soil should be studied by digging a trench 3 feet deep, 6 feet long, and at least 2 feet wide. Observations should be made for mottling in the A_1, A_2, and A_3 horizons, or layers (see Figures 5–11, pages 240 ff.), root growth, plow sole, hardpan, and signs of good aeration in the different layers. Soil samples should be taken in each horizon.

STEP 2. Determine the percentage of the base saturation with calcium, because research work done by soil colloid chemists indicates that 85 per cent of the base exchange in the soil to a depth of three or more feet must be saturated with the calcium ion before maximum yields can be expected. The base saturation must be determined by a calcium test rather than a soil acidity test. The acidity test does not differentiate between calcium and such other ions as potassium, magnesium, sodium, and ammonium. If the acidity test is used, we never do apply sufficient limestone to reach the necessary 85 per cent calcium saturation.

STEP 3. Consider tillage methods—whether subsoiling or other practices are necessary. Consider minimum tillage—decide on distance between plants.

STEP 4. Plant crops with fertilizer solutions. Not more than four gallons of 10–20–10 or its equivalent.

STEP 5. Spray foliage with 10–20–10 fertilizer solution.

Index